EARLY LATHES & MACHINE TOOLS OF INTEREST

VOLUME 1

by C.L. Deith

0 905100 45 X

The articles entitled 'Early Lathes and Machine Tools of Interest' which have appeared in the monthly magazine 'Engineering in Miniature' have attracted considerable interest. In addition to making fascinating reading the series provides information for any readers who may have the opportunity to buy, secondhand, one of the lathes featured with which they would not normally have been familiar.

It was whilst looking through early technical magazines for information to complete one of these articles that I realised the interest and information that could be gained just from the advertisements themselves as published by various manufacturers. Indeed numerous lathes were advertised in the early years of this century which are now quite unknown to us, for example, whilst everyone knows of the existence of the Drummond and Exe Lathes, Christopher's Combination Machine patented in 1906 is as you'll see from the manufacturers advertisement, a fascinating looking piece of equipment.

I have therefore taken the opportunity to collect together a selection of the most interesting advertisements covering the period 1900 to 1930 which feature many of the lathes and other machines as well as some of the products advertised over the period. I have endeavoured in the majority of cases to use advertisements actually containing an illustration of the product and from which one can see many details of the machines. As previously mentioned the collection contains a number of machines which have received very little notice and are generally unfamiliar to us such as the Ashby Centre Lathe, Airde & Company's Wade Lathe, which has certain similarities with the Drummonds and the Nicholls Brothers NW Lathe. I have also taken the opportunity to reproduce some advertisements of the American manufacturers such as Star and Shepherd.

The Drummond advertisement showing the light car built on their 3½" Lathe dates from 1918. This is one of the first advertisements actually featuring something made on a lathe although Drummonds did make quite a regular practice of featuring the models built on their products later on in the 1930's. It is interesting to note that this practice has recently been re-introduced by Myfords Ltd who now feature some of the excellent models made at the present time on their machines in their advertising. One doubts, however, whether it would be possible to build a car in this manner today and yet still ensure the vehicle satisfied the many numerous requirements of the Road Traffic Act.

I trust that readers will enjoy this collection of facsimile advertisements. It is hoped to issue a further volume on Lathes and Machine Tools as well as other volumes dealing with castings and part completed models offered by the trade to early model engineers. A further volume is to be issued covering gauge 0 and gauge 1 model railways and other models.

July 1982

C L Deith
Hinckley, Leics.

"THE BEST VALUE I HAVE EVER SEEN."

Unsolicited fiom a Working Mechanical Engineer.

"I CONSIDER IT A SPLENDID BARGAIN."

Unsolicited from a Experienced Amateur.

PEDESTAL

SCREW-CUTTING LATHE

HOLLOW SPINDLE,

$\frac{29}{32}$-in. bore.

Centre Height, $4\frac{3}{4}$ ins.

Swing in Gap, 16 ins.

All Change Wheels Machine Cut.

Ball-bearing Thrust.

Back-geared.

Total Length, 3 ft. 6 ins. Between centres, 1 ft. 8 ins.

Price, including Set of Change-Wheels, Compound Slide-rest (graduated), Foot Gear, Driver Plate, Centres and Spanners, and Self-acting Cross-feed.

£22 10s.

Send 3d. in Stamps for Illustrated Price List of Lathes, Shaping, Milling, and Drilling Machines.

GEORGE ADAMS,

144, High Holborn, London, W.C.

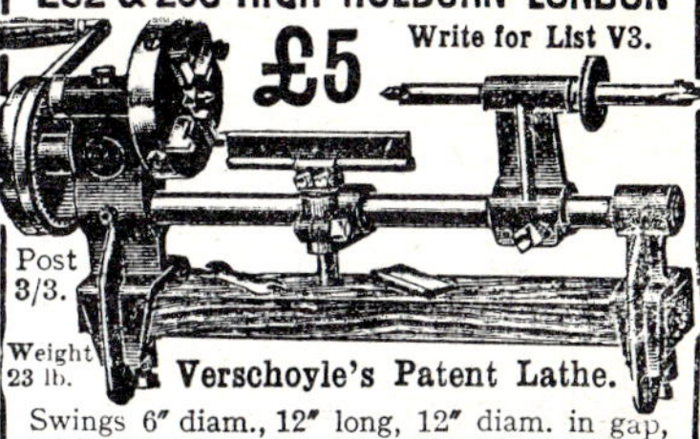

Swings 6″ diam., 12″ long, 12″ diam. in gap, with 5″ four-jaw chuck complete £5, post 3/3. Set of six handled Woodturning Tools, 9/- extra. **Complete List Woodworkers' Tools, No. M291, post free 2d.**

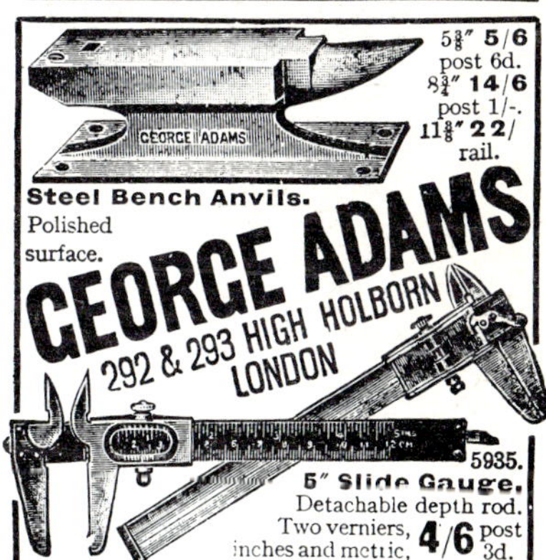

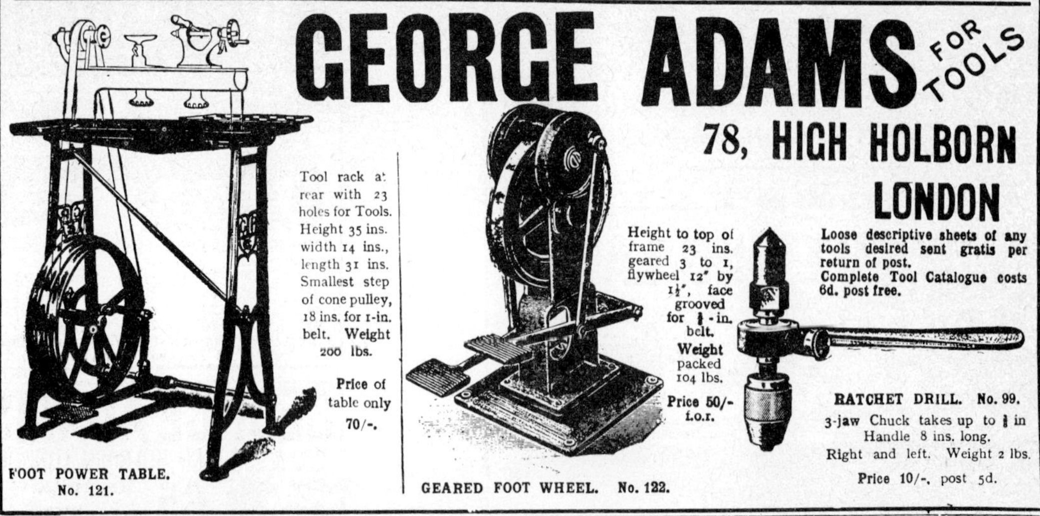

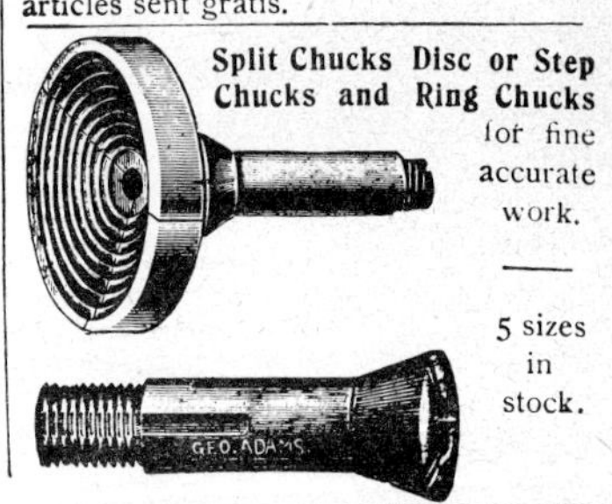

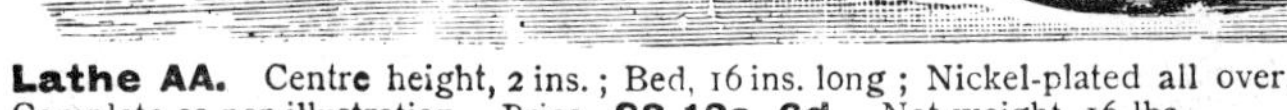

Lathe AA. Centre height, 2 ins.; Bed, 16 ins. long; Nickel-plated all over. Complete as per illustration. Price, **£8 12s. 6d.** Net weight, 16 lbs.

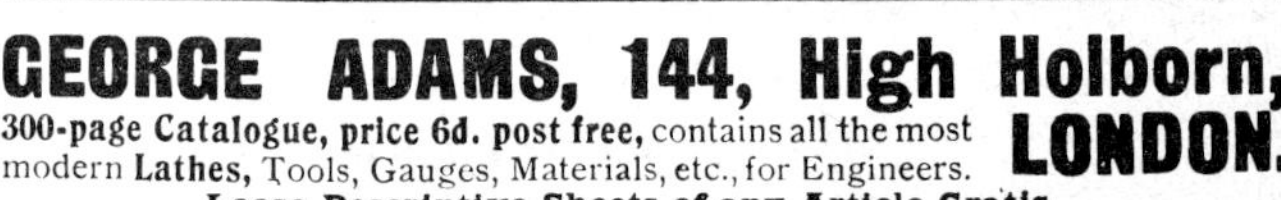

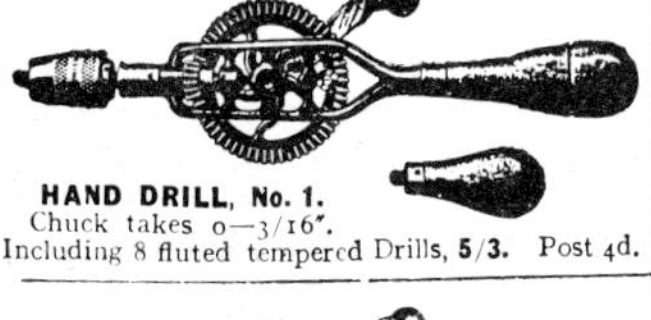

HAND DRILL, No. 1.
Chuck takes 0—3/16".
Including 8 fluted tempered Drills, **5/3.** Post 4d.

HAND DRILL, No. 5.
As No. 1, but with extra wide hand wheel for use in delicate work, with 8 Drills. Price **6/3.** Post 4d.

UNIVERSAL HANDLE, 5" long.
Of malleable iron, for all kinds of files and small tools. Price **6d.** Post 2d.

SCREW-CUTTING LATHE, "A." Centre height 5½", swings in gap 16½", bore of spindle 29/32", 20" between centres. Price **£27 15s. 0d.**

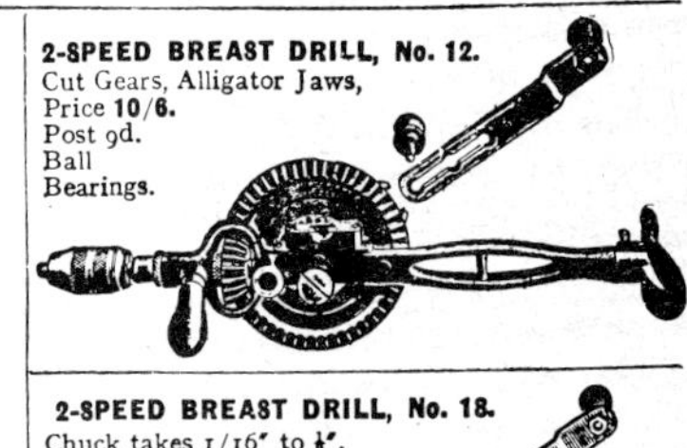

2-SPEED BREAST DRILL, No. 12.
Cut Gears, Alligator Jaws,
Price **10/6.**
Post 9d.
Ball
Bearings.

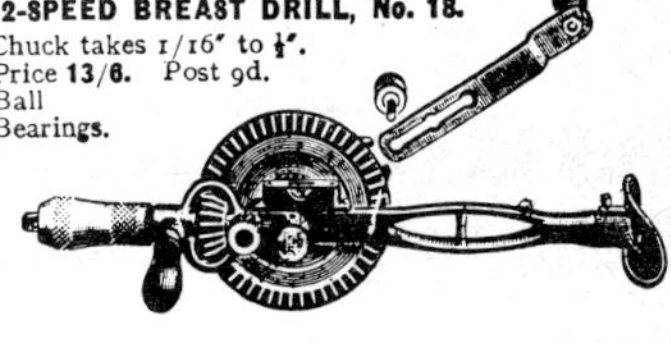

2-SPEED BREAST DRILL, No. 18.
Chuck takes 1/16" to ½".
Price **13/6.** Post 9d.
Ball
Bearings.

The Engine you have been looking for!

The AMANCO MIDGET starts instantly by means of a self-winding strap, and runs without any attention beyond filling with petrol and oil. It is ideal for running small lathes, etc. Takes up no room, and can be mounted on the machine itself. Write for list "M.E."

ASSOCIATED MANUFACTURERS' Co. (London), Ltd.
46–48, Wharfdale Road, King's Cross, LONDON, N.1.

"ASHBY JUNIOR" LATHE

$2\frac{1}{2}$" centre.

Price £2 : 0 : 0 as plain lathe
£2 : 10 : 0 with Compound Slide Rest.

Further particulars—send stamp to—
ASHBY ENGINEERING CO.,
ASHBY ROAD, BROCKLEY, LONDON, S.E.4.

Ashby 3" Centre Lathe.

All steel bed, hollow mandrel, will take $\frac{3}{8}$" bar
PRICE £2 5s.,
with T-rest, driving plate.
For further particulars send stamp to—
ASHBY ENGINEERING CO.,
Ashby Road, Brockley, S.E.4.

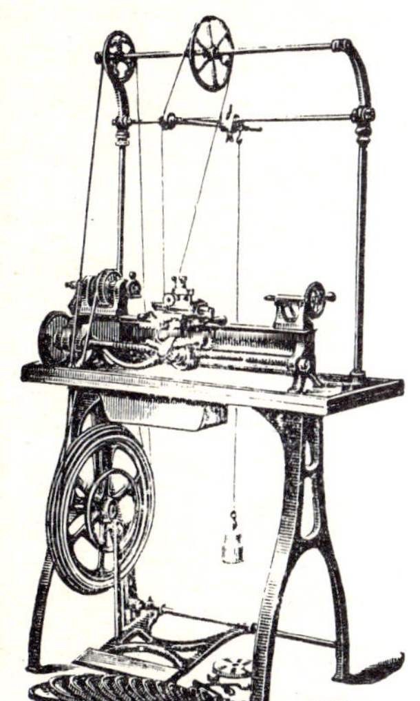

NOTICE.

WE ARE BONA FIDE MAKERS OF OVER

250 VARIETIES OF

LATHES

From the Bench Lathe at 35/- to the Engineers' Lathe at £250.

PLANERS, SHAPERS, DRILLS.

SEND DETAILS OF YOUR REQUIREMENTS.

CASH OR HIRE PURCHASE!!

Call at London Depôt—100 Houndsditch. All Letters to Colchester.

BRITANNIA CO., COLCHESTER,

ENGLAND.

BRITANNIA LATHES and Modern Machine Tools

Buy a First-class Lathe at a Moderate Price.

Have supplied thousands of Lathes and Machine Tools to the Admiralty, War Office, India Office, etc., etc.

No. 16 Lathe.

Write for Catalogue, and Specify your Requirements.

ALSO MAKERS

Britannia Safety - -

Automatic Oil Engine,

THE MOST PERFECT POWER FOR DRIVING SMALL WORKSHOPS.

HEAD OFFICE & WORKS:

BRITANNIA ENGINEERING Co., Ltd., COLCHESTER

BRITANNIA $4\frac{1}{2}$ Inch

MILLING, GRINDING. GEAR-CUTTING, BORING, SELF-SURFACING, SCREW-CUTTING

LATHE.

Embodying every possible improvement in design, including **ROLLER-BEARING HEADSTOCK,** full **BALL-BEARING TREADLE** motion and **VEE BED.**

Also embracing the novel and valuable feature that at some future date you can purchase the necessary parts, at moderate cost, and fit same to your standard Lathe to obtain: **AUTOMATIC FEED TO TOP SLIDE,** or convert it into a **MOTOR-DRIVEN LATHE** as illustration.

We are selling a limited number of these Lathes on

EASY PAYMENT TERMS.

£8 : 10 : 0

DEPOSIT

and the balance spread over a long period.

DELIVERY FROM STOCK.

Many Improvements.

Standard Lathe,

£34 : 0 : 0

Send for Catalogue.

BRITANNIA LATHE & OIL ENGINE Co., Ltd., Colchester.

No. 3, Backgeared Lathe.
3 ins. Centres, 2 ft. 6 ins. Bed.

Price as shown, £6.
With Stand, Treadle, and Flywheel, £6 10s.

No. 5, Backgeared Lathe.
4 ins. Centre, 4 ft. Bed.

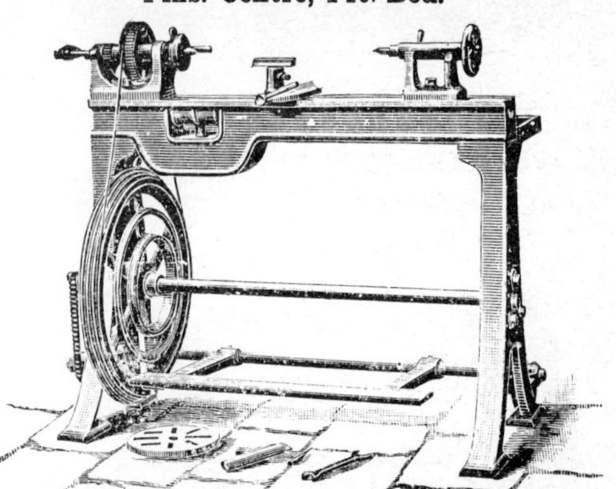

Price as shown, £12 15s.

No. 14, Screw-cutting Lathe.
3½ ins. Centres, 3 ft. 6 ins. Bed.

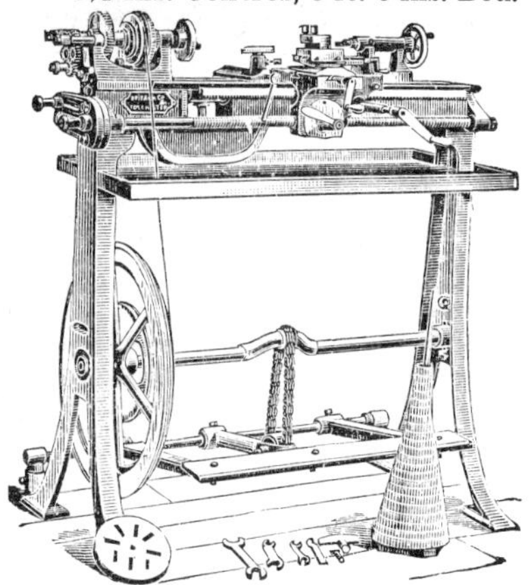

Price as shown, £18 18s.

No. 15, Screw-cutting Lathe.
From 4 to 5 ins. Centres, 4 to 5 ft. Bed.

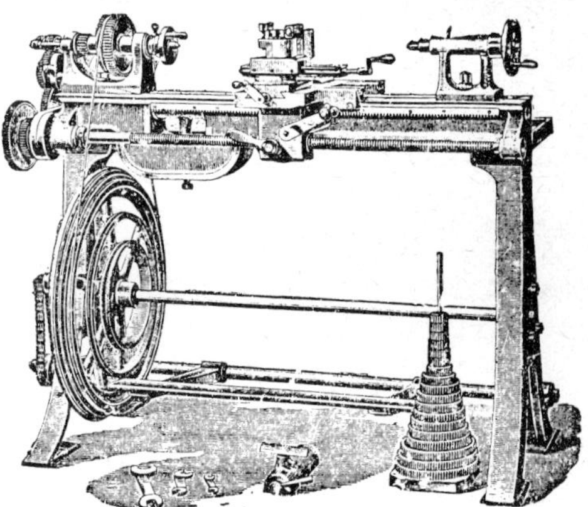

Price of 4-in. by 4 ft., £25 4s.
Price of 5-in. by 4 ft., £28 7s.

BRITANNIA CO.

MAKERS OF
250 Varieties of LATHES, SHAPERS, DRILLS, &c., &c.

NOTE!—Most of our Lathes are made with steel-wearing parts, and cut gears in head.

WORKS—COLCHESTER.

All Letters to Britannia Works, Colchester.
LIST OF SECONDHAND MACHINERY, ENGINES, etc., 2d.

London Warehouse:
100 HOUNDSDITCH.

Send 1s. for New Catalogue, which amount is returned if Goods to the value of 21s. be ordered.

Contractors to the ADMIRALTY, WAR OFFICE, and INDIA OFFICE. Numerous Prize Medals.

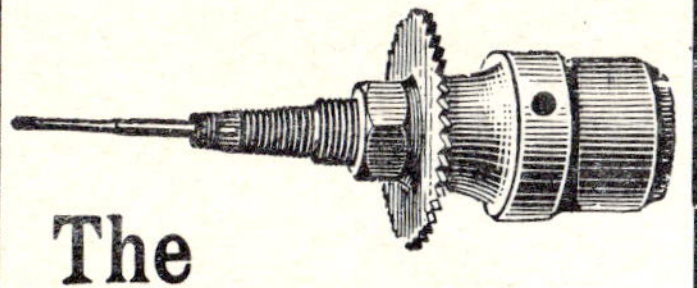

The Jewellers' Chuck

Will carry Drill, Circular Saws, Bobs, and Polishing Brushes. Made with 5/8-in. Shank to fit any Chuck. Price **5/-** each. Saws, **2/6;** Polishing Brush, **1/6;** Emery Wheels, **2/-**

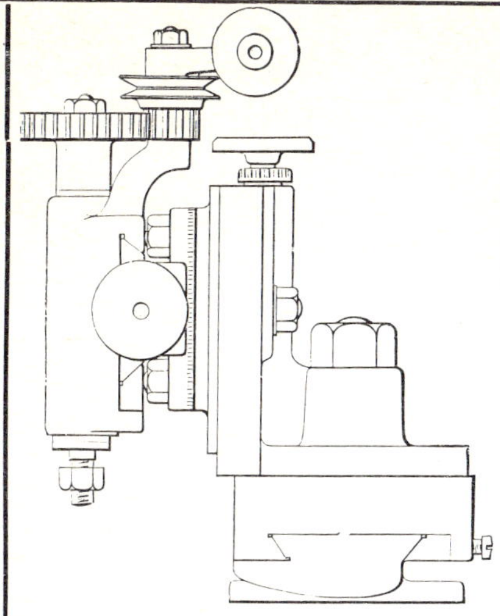

Vertical Milling and Fluting Attachment for Lathes—

for Fluting Taps, Cutting Gear Wheels and Racks. Any sort of gear can be accurately cut with a suitable division plate on Lathe head. The Milling Spindle works vertical or any other position.

Send for Circular. 

Circular Saw and Platform for Lathes

(any Size).

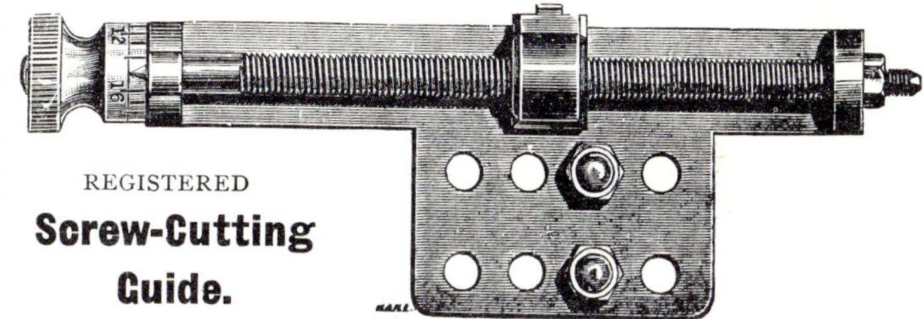

REGISTERED

Screw-Cutting Guide.

For 3 ins. Lathe ..	18/-	For 5 to 6 ins. Lathe ..	20/-
For 4 ins. Lathe ..	18/-	For 8 to 10 ins. Lathe ..	22/-

The above appliances can be used on any Lathe. By its use the workman can regulate the depth of cut to the greatest nicety, and the use of the chalk mark, or any such expedient, unnecessary.

It can be used for inside or outside screw cutting, or other work requiring uniformity. It saves time from insufficiency of cut. It prevents the breaking of tools, or the work being torn out from the centres. It is a reliable stop for ornamental drilling and fluting.

While the Lathe is cutting, this can be adjusted for the following cut. It only requires to be bolted upon the saddle of a Screw-cutting Lathe, and a projecting stud or screw fixed in the middle engages the stop.

It can also be used on ordinary Lathes, with slide-rest—in this case it must be fixed to the bed.

This is a tool which has long been wanted by engineers, and will also be appreciated by amateurs.

The "Climax" Tool Holder.

SIDE RAKE AS WELL AS TOP RAKE.

The Most Perfect Tool Holder yet Invented.

Makers of every description of Chucks and Appliances for Lathes.

BRITANNIA COMPANY, Colchester,

Makers of Engineers' Tools—250 Varieties.

SEND 1/- for NEW CATALOGUE—Allowed if Goods to 21/- be Purchased.

FOR NEW DESIGNS

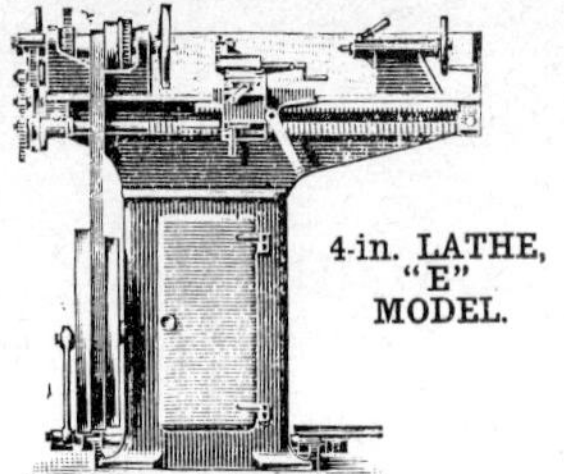

4-in. LATHE, "E" MODEL.

in small High-class

Screw - Cutting Lathes.

W. H. ASTBURY, Wheeler Gate, NOTTINGHAM.

☞ UNPRECEDENTED OFFER! — We will furnish **Castings and Parts**, including full-size **Working Drawings** of above excellent Lathe, to *bonâ fide* Amateurs and Mechanics to fit up for their own use. Send Stamp for detailed List.

3" **Lever Scroll Chucks,** with two sets of jaws, **24/-. Backplate Castings, 1/-;** or finished to suit your mandrel, **5/-,** post 1/-. Best British material and workmanship. Ideal for your small lathe. Lists Free.

The Ideal Toolholder for parting-off, screw cutting, etc. 5 sizes. On appro. if desired. Prices include post.

Lists Free.

No. 0 for lathes 2½"-3" **10/6,** extra cutters **9d.**
No. 1a „ „ 3½"-4½" **15/9,** „ „ **1/-**
No. 2a „ „ 5"-6½" **21/-,** „ „ **1/6**

Swivelling Toolholders save time and cutters. Cutter Swivels through over 180°. 7 sizes.

Prices include cutter and post.

No. S1 for lathes 2½"-3" **5/6,** extra cutters **6d.**
No. S2 „ „ 3½"-4" **10/-** „ „ **6d.**
No. S3 „ „ 4½"-5" **11/7,** „ „ **8d.**
No. S4 „ „ 5½"-6" **13/3** „ „ **1/3**

Also High-Speed Cutters and Cutting-off Blades for all makes of Toolholders.

Why not Write Us To-day?

F. Burnerd & Co. (Dept. M),
DRYBURGH WORKS, DRYBURGH ROAD, PUTNEY, S.W.15.

3" **Lever Scroll Chucks,** with two sets of jaws, **24/-. Backplate Castings, 1/-;** or finished to suit your mandrel, **5/-,** post 1/-. Best British material and workmanship. Ideal for your small lathe. *Lists Free.*

F. BURNERD & CO. (Dept. M), Dryburgh Wks., Dryburgh Rd., Putney, S.W.15.

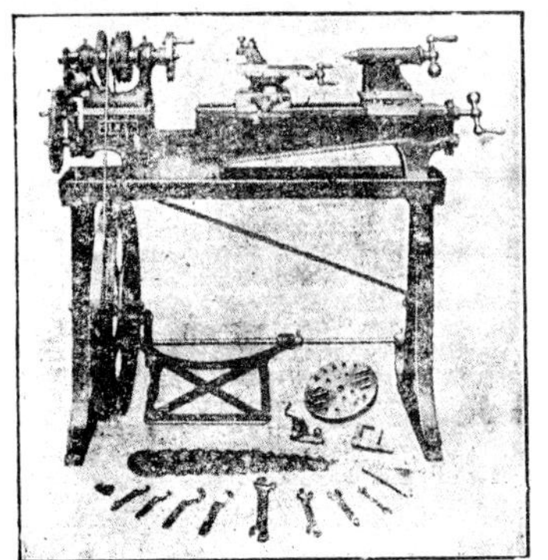

The "I.X.L.," 5⅛-in. centre.
£16 : 0 : 0.

All Orders despatched same day as received.

LATHES.

"I.X.L.," "J.B.," DRUMMOND, "PATRICK," "RELMAC," etc.

We have them All in Stock, ready for Immediate Delivery.

LET US QUOTE YOU for your Lathe;. fitted with Chucks, Tools, etc., complete.

If you have not yet decided on your particular Lathe send for Lathe Sheets, free.

COMPLETE CATALOGUE FREE BY POST, 5d.

J. BUCK, 56, Holborn Viaduct, E.C.

BUCK & RYAN

TOOL MERCHANTS,

310/312, Euston Rd., London, N.W.1.

Ideal Gifts for The New Year for Father or Boy

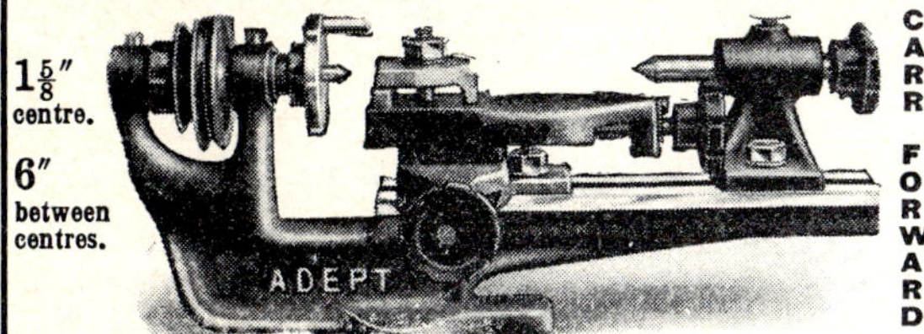

The "ADEPT" Lathe with Compound Slide-rest, **20/-**
With Screw Tailstock, 5/- extra.

FOOT MOTOR FOR ABOVE, 20/-

ALL OTHER ACCESSORIES KEPT IN STOCK.

The "PORTASS" Lathe, Model B, **55/-**
Model de Luxe, **85/-**

ALL PORTASS LATHES IN STOCK.

BE SURE TO VISIT STAND 27

SHOWING

PORTASS LATHES

BUCK & RYAN

FROM £1	"MODEL ENGINEER" EXHIBITION LONDON	TO £16

THE "RAPID"

Hardening & Tempering Stove.

Suitable for all kinds of Small Tools such as TAPS, DIES, DRILLS, REAMERS, Etc., Etc.

Cost of running ½d. per hour, and a temperature of 900 degrees Centigrade can be obtained in the lower chamber.

MADE IN TWO SIZES.

No. 1.	Capacity $3\frac{1}{2}'' \times 2 \times 2''$	..	40/-
No. 2.	,, $6\frac{3}{4}'' \times 2 \times 2''$	..	60/-

For casehardening in above Stove use our **Kasenit.**
The purest & best compound on the market

For further particulars write to:

C. W. BURTON, GRIFFITHS & CO.,
Ludgate Square, LONDON, E.C.

SUPER RELM 4½" S.S. S.C. LATHE, Gap Bed, Hollow Mandrel 1¼" × ⅞" Bore.

Relmac and Super Relm Lathes.

PRICES AGAIN REDUCED.

We are always endeavouring to cheapen cost of production and to reduce our prices so that we can pass on the benefit to you.
Our Lathes are RELIABLE, EFFICIENT and ACCURATE.
You can trust both the Machine and the Maker.
Our catalogue and list of reduced prices is yours for the asking.

Have you seen the RELMIL MILLING ATTACHMENT

CHELTENHAM WORKS, Ltd.,

'Phone: Brixton 2001. **86, ACRE LANE, BRIXTON S.W.2.**

MODEL ENGINEER EXHIBITION.

Cheltenham Works, Ltd.,

86, ACRE LANE, BRIXTON, S.W.2.

At Stand No. 20.

The Makers of RELMAC and SUPER RELM Lathes beg to present their New Machine,

The "RELMINOR"

a Lathe with Gap Bed, 3" Centres, 12" between Centres, Hollow Mandrel ⅝" Bore;

A STURDY YOUNG ARRIVAL

Fully capable of sustaining the enviable reputation associated with the RELM Series of Machines.

A Lathe for Serious Work at a Low Price,

The result of months of designing and experimenting with the object of reducing price whilst yet producing a machine which can seriously be described as

A LATHE!

Come and see this Machine at our Stand, where you will also find various types of the famous RELMAC and SUPER RELM Lathes, together with many useful and unique attachments and accessories, including a bracket which will convert your Relmac Lathe into a Vertical Milling Machine at a cost of only 18/6.

This is Your Opportunity to make an examination of our Machines and see them at Work. Our representatives are at your disposal for any information you require. R.S.V.P.

A CHRISTMAS PRESENT WORTH SENDING.

The Most Popular 3-in. Lathe on the Market.

Because we have had a large demand for this size Lathe alone; sold them not only in England, but in other countries of the World.
We sell one Lathe; it sells another, and so it goes. Why? Because the Lathe is made right, and it pleases the buyer. Our Lathes are the most rapid working, because they are the most convenient to operate; strongest, because they are correctly designed; most durable and accurate, because the material and workmanship are the best.

THIS CUT REPRESENTS OUR 3-in. CENTRE CHALLENGE LATHE.

Price List of Finished and Unfinished Castings, 2 stamps.

Maker: W. H. CLARK, Engineer & Tool Maker, Junction, Windhill, Shipley.

The "Challenge" Lathe

3-in. Centres.

17/6

3-in. Centres.

PRICE LIST OF FINISHED & UNFINISHED CASTINGS 1d. STAMP.

Maker **W. H. CLARK** Engineer and Tool Maker,
JUNCTION, WINDHILL, SHIPLEY, YORKS.

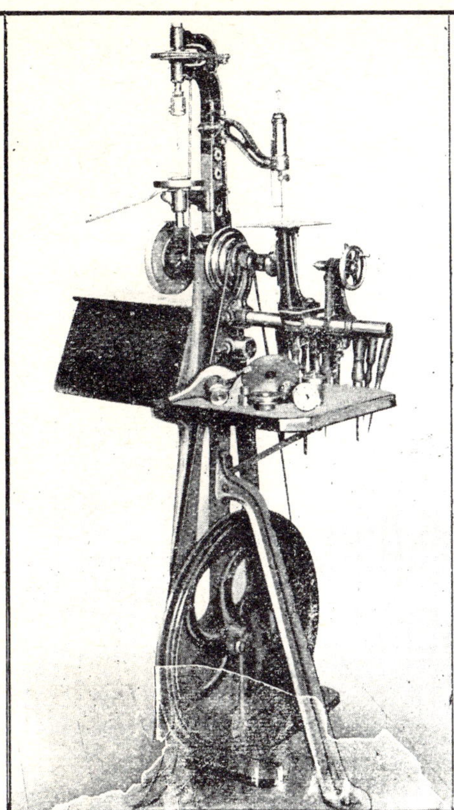

CHRISTOPHER'S

COMBINATION MACHINE

(Patent April, 1906),

Lathe, Drill, Jig Saw, Emery Grinder, Vice, &c., &c.

Lathe, taking up to 30-in. swing for Pattern Making, **£5 10s.**

EXTRAS—

Jig Saw, taking 2-in. wood.
Sensitive Drilling Machine
4 Speeds, 4 Heights.
Emery Grinder.
Tool Cabinet.
Vice and Back Gear.

Complete Machine, £10 10s.

J. CHRISTOPHER & SONS, Engineers' Stores,
39, 41, & 43, Clerkenwell Rd., London, E.C.

Teleph·ne: 1948 Holborn.
Lists, 2d. postage.

TREADLE LATHES from 90/-
Parallel Vices from 5/3.
Sliding, Surfacing and Screw cutting Lathes.
ANGLE & FACE PLATES. CHUCKS & DRILLS
Leather Belting Lubricating Oils.
GAS ENGINES & MOTORS.
J. CHRISTOPHER & SONS,
35 & 35B, Clerkenwell Road, London, E.C.

DIGNUS ENGINEERING Co.,
ORLEANS ROAD,
TWICKENHAM, S.W.

3" Screwcutting Lathe.
Stamp for particulars.

DIGNUS

SCREW-CUTTING LATHES.

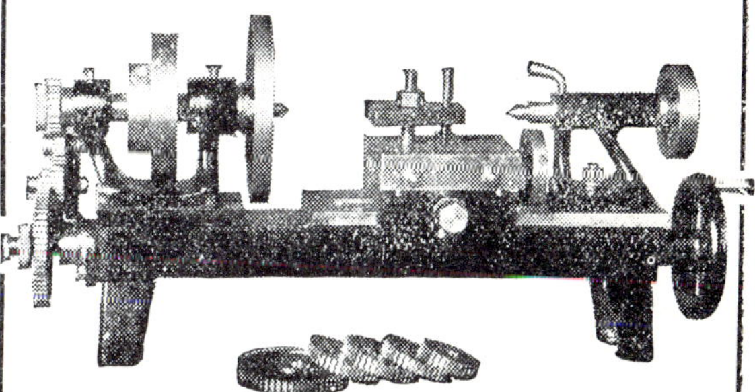

3-in. CENTRE BENCH LATHE.

19-in. Bed. 9 ins. between Centres.
29-in. Bed. 19 „ „ „

ALSO BUILT WITH TOTALLY ENCLOSED REDUCTION GEARS IN CONE PULLEY.

Although not exhibiting at the "M.E." Exhibition our Lathes can always be inspected at 10, Grays Inn Road, W.C.2.

Dignus Engineering Co.,
Orleans Road, Twickenham, S.W.

NEW HAND-LEVER BENCH SHAPING MACHINE (Shown Cutting Keyway in Shaft passing through Body of Machine).

Drummond Bros.' New Machine Tools.

SPECIALLY DESIGNED FOR MODEL WORK.

MODEL-MAKING having now become a Fine Art, the old-fashioned, weak, and inaccurate Tools are almost useless. We have endeavoured to design a Lathe and Shaper which, though costing less than the old-fashioned, clumsy Tools, are capable of the finest and most accurate work that can be desired, and of power sufficient to deal with cast iron and steel work in a proper manner.

Our method has been to design the Tool from the beginning, unhampered by having to use old-fashioned stock-patterns, and to manufacture it when designed by the very latest methods and by the use of the latest automatic machinery. This enables us to turn out a highly-finished, perfectly accurate Tool at a price which compares favourably with that charged for common, out-of-date articles. Our Book on the Lathe, advertised on another page, will prove to intending customers the utility of our new patented designs.

Short Specification of New Patent Gap-Bed Lathes, Self-Acting Sliding and Screw-Cutting.

Back-geared headstock; gear machine cut from solid blanks; 3-speed cone for flat belt; double-cone mandrel, new design, running in hard gunmetal bushes; poppett head, with long bearing barrel, fitted with steel square thread screw; extra strong sliding saddle, with recessed cross slide and compound slide-rest, swivelling completely round; steel square thread draw screws, accurately cut; 4-pitch steel square thread lead screw; patent treadle motion, running in anti-friction metal bushed swivel bearings; accessories; truly-surfaced ⊥ slotted boring plate for cylinder boring, &c.; face-plate; driver plate; travelling steady with 3-jaws (hardened steel); hand-rest, with two tees; hardened tool steel centres; driving belt; and spanners.

Price	..	**4-in. Centre, 3 ft. 6-in. Bed**	..	**£18 10 0**
„	..	**5-in. Centre, 4 ft. 6-in. Bed**	..	**£25 0 0**

Delivered Guilford Station (L.S.-W. Railway), Box Returnable.

Short Specification of New Hand Lever Bench Shaping Machine, with Self-Acting Feed.

This machine is practically a reduced copy of the best type of power shaping machines. It works easily and with absolute accuracy; takes a good stiff cut, and leaves a highly-finished, truly flat and straight surface. It will accurately face cylinder ports, faces slide valves, engine beds, &c., besides cutting key-ways (in any length shaft), fluting-taps, and rimers, and any other shaping work.

Price, including Swivelling Machine, Vice (graduated), Three Tools, Spanners, &c. £7 10s., delivered Guildford (L.S.-W. Railway), Box returnable.

SPECIAL NOTICE.—Although we make our tools to order, we will *in every case send same on approval for ten days.* We have now had this system in operation for two years, and have as yet had no tool returned, but have instead received from *every one of our customers a good testimonial.* These can be seen at our office.

Drummond Bros., Engineers, Pink's Hill, nr. Guildford, Surrey.

THE LATHE THAT COST £2,500

SOME SPECIAL ADVANTAGES of the NEW DESIGN 3½ in. centre, 2 ft. 6 in. bed, self-acting, boring, sliding, and screw-cutting MODEL MAKERS' LATHE, made by DRUMMOND BROS., Ltd.

THIS is the tool which we lately advertised as having cost £2,500, that being the amount spent on quite special machinery before the first lathe was produced. This statement may appear to some as merely a catching headline, but it was a very real fact to us, and also to our customers, who, owing to the method of manufacture (although involving a heavy first cost) giving rapid and extremely accurate production, can obtain a lathe of at least £30 value for £13 10s.

We intend to give fortnightly on this page an illustration of this lathe doing various jobs, which, except on such a specially designed tool, would be difficult. These will interest you, we hope, whether you are a prospective purchaser or not.

BORING A $2\frac{3}{4}$ H.-P. DE DION PATTERN MOTOR CYLINDER; BORE, 75 MM. (ABOUT 3 INS.).

Observe the ease with which such a cylinder (usually a very awkward thing to fix to a small lathe) can be bolted to the self-acting boring carriage, which, in this new design, forms the saddle for the slide-rest. The slide-rest can be seen below the bed, on the tray, having been entirely removed by slacking one nut, to leave the flat-surfaced L slotted plate which is just the very thing to bolt work to for boring.

The photo does not show, as it should, that the boring plate is fitted to the Vees of bed and adjustable for wear. In the photo, the cut being taken is a finishing one, the change wheels at headstock being set to give a cut of 110 revolutions of cutter per inch of travel; this with a proper cutter gives very perfect results. Fast returns between cuts are made by handwheel on screw. This handwheel is useful also for roughing or trial cuts. Lathe is perfectly made, highly finished, perfectly accurate and carries our guarantee, and is sent on approval for 10 days' trial.

Price £13 10s. complete with treadle; or as Bench Lathe, £10 10s.

Less 5 per cent. cash discount, or in twelve monthly payments.

WRITE FOR PARTICULARS TO

DRUMMOND BROTHERS, Ltd., RYDE'S HILL, nr. GUILDFORD.

THE NEW DRUMMOND FIVE POUND LATHE.

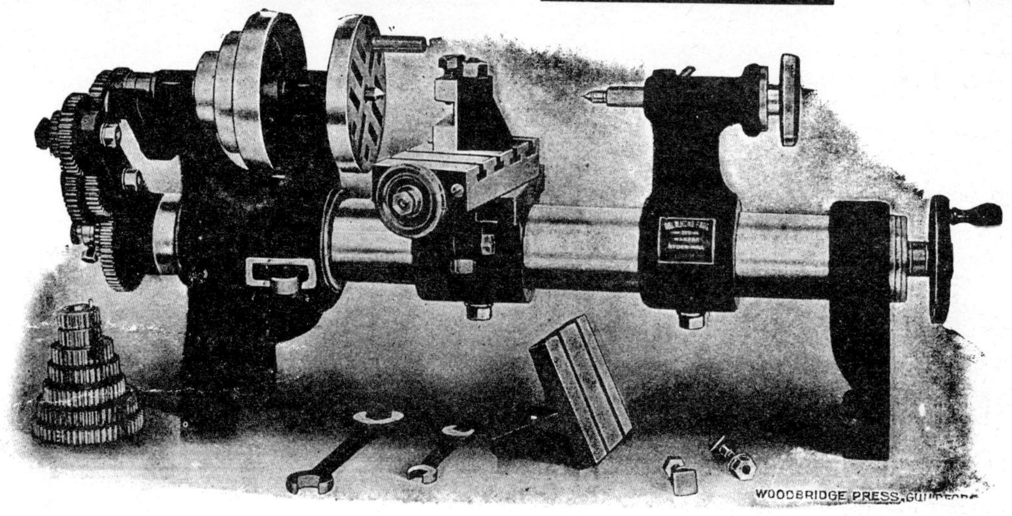

A Self-acting, Sliding, Boring, Screw-cutting and Milling Lathe.

Write for particulars to

DRUMMOND BROTHERS, Ltd., Rydes Hill, near GUILDFORD.

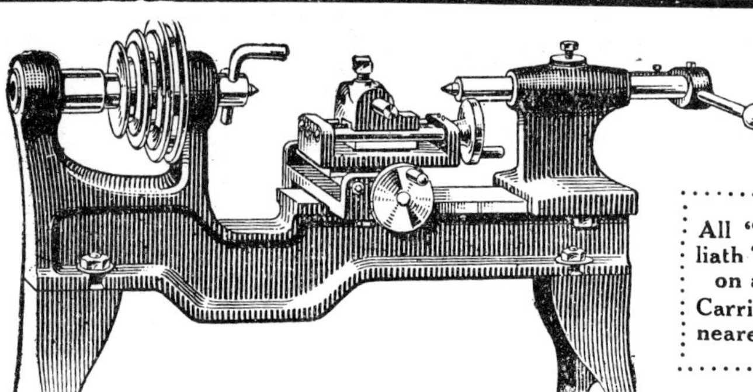

The ideal Xmas Gift for Model Engineers.

All "Little Goliath" parts sent on approval. Carriage paid to nearest station.

The "Three-in-One" Machine.

THE LITTLE GOLIATH, the small lathe with the compound slide rest, is a splendid little tool for the model engineer, but its utility may be greatly increased by the new patented conversion sets, which enable the lathe to be used as vertical mill and vertical drill.

Lathe, drill, and mill may all be driven by a sewing machine treadle, or by the proper footmotor as shown here.

Price of Lathe alone ...	£3 15 0
Both Sets of Parts, for conversion to Mill and Drill	£1 19 0

Write for list and details of deferred payment system.

DRUMMOND BROS., Ltd., Rise Hill, Guildford.

"You turn round to look at DRUMMOND LATHES."

The illustration here shows the 4-in. Lathe mounted on Stand.

Price, Complete, £8

Or as Bench Lathe only, £5.

Full Particulars gladly sent to any Address.

DRUMMOND BROS., Ltd.,

RYDE'S HILL, :: ::

GUILDFORD, SURREY.

This Efficient Light Car

was constructed in an amateur workshop—the chief metal-working tool being a

3½-in. Drummond Lathe

On it all turning, screw-cutting, boring, milling, and drilling was done.

The constructor, Mr. W. Tayler, of East Sheen, in an article in *The Motor* describes how he overcame all his difficulties by the use of the well-known Drummond Lathe

Drummond Bros.
LTD.,
Rise Hill, Guildford.

Drummond Brothers', Ltd., New Design, 3½ ins. centre, Self-Acting Sliding, Boring, and Screw-cutting Lathe.

THIS fine little tool has lately been carefully gone over, to see where improvements were possible, and the improved tool for 1906 is now ready for delivery. The most important improvements are a stiffening up throughout, stay-piece under tailstock preventing any springing in of bed when locking.

The Headstock Design is now as shown (taken to pieces).

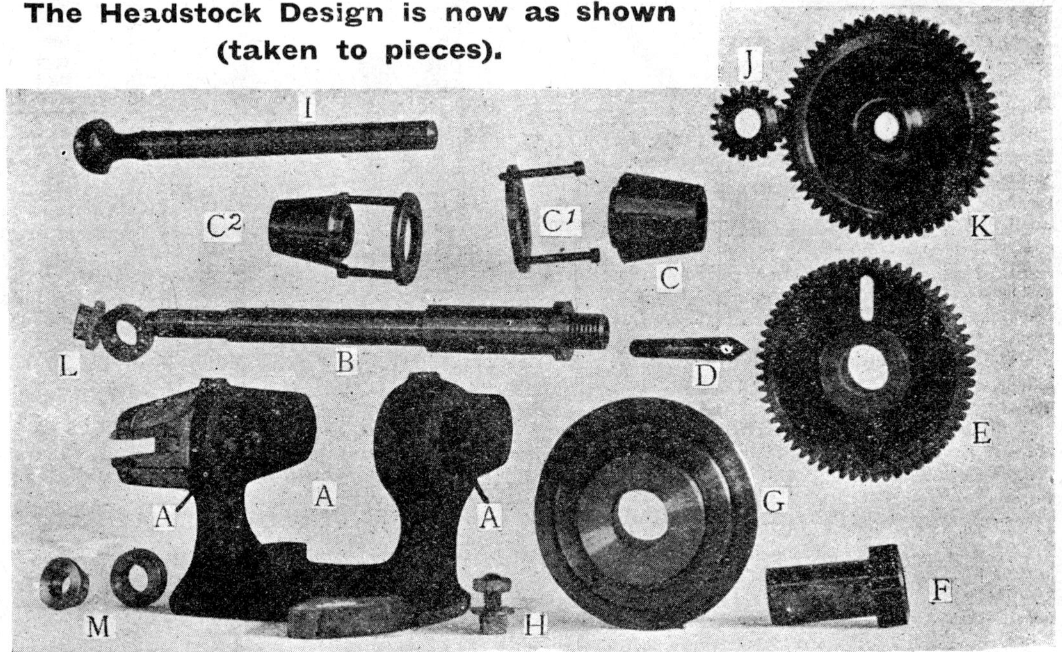

A.—Head casting, showing taper bore and interior oil recesses A A.
B.—Hollow mandrel, parallel bearings.
C.—Front gunmetal bearing, split and coned exteriorally.
C1.—Steel ring and adjusting screws.
C2.—Rear gunmetal cone, ring and screws.
D.—Tool steel centre, No. 1 Morse taper socket.
E.—Driving gear wheel, fixed to mandrel.
G.—3-speed **V**-grooved pulley, fast on quill sleeve.
H.—Disengaging lock between E and G.
I.J.K.—Back gear spindle, pinion and wheel.
L.M.—Mandrel locking and adjusting collars.

THIS TYPE OF MANDREL and Adjustment will be recognised as being very similar to that adopted by Messrs. Brown & Sharpe, and other high-class makers on their latest Milling Machine spindles, and we can guarantee that this is about the finest possible form for bearings for wear and for retaining accuracy; the take-up can only be truly concentric, as a study of the forces at work when adjusting will show

NO INCREASE IN PRICE.

Short Specification.—3½ ins. centre set over back-geared headstock, steel hollow mandrel, set-over tailstock, 2 ft. 6 ins. gap bed, steel square thread lead screw, **T**-slotted boring carriage (carrying slide-rest), steel square thread draw screw, change wheels and all gears machined cut from solid, all slides hand-scraped to perfect fit, accuracy throughout guaranteed.

Price complete with Treadle, **£13 10s.,** or as Bench Lathe **£10 10s.**

Write for Lists and particulars of Approval System to—

DRUMMOND BROTHERS, LIMITED,

Ryde's Hill, nr. Guildford, Surrey.

DRUMMOND BROS.' 5-INCH CENTRE SELF-ACTING SLIDING AND SCREW-CUTTING LATHE.

For the Larger Workshops

Like the smaller sizes of Drummond Lathes, the larger models are known and used practically all over the world. The type illustrated, made in 5″, 6″, and 7″ sizes, forms the ideal tool for the amateur worker whose requirements call for something larger than the famous "3½"." The gap bed is of a special section designed to minimise wear. Fine-limit accuracy is guaranteed. Treadle drive models can be supplied, and a full range of attachments is listed.

Write for illustrated catalogues, instalment terms, etc.

DRUMMOND BROS., LTD.

Rise Hill, GUILDFORD.

Motor Drive 3½ in. Lathes

A handy arrangement with treadle for occasional use

If you have electricity in the house, a motor driven lathe is a great boon. The arrangement shown does away with the necessity for shafting, motor base or any extra fixtures, and is self-contained, rigid, and readily installed. Full details, sizes, etc., in our booklet. Quotation on receipt of particulars of current available. Write:

DRUMMOND BROS. LTD.

Rise Hill, Guildford.

Typical jobs on a DRUMMOND 4 in.

Two examples of convenient machining jobs on the Drummond 4 in. Lathe.

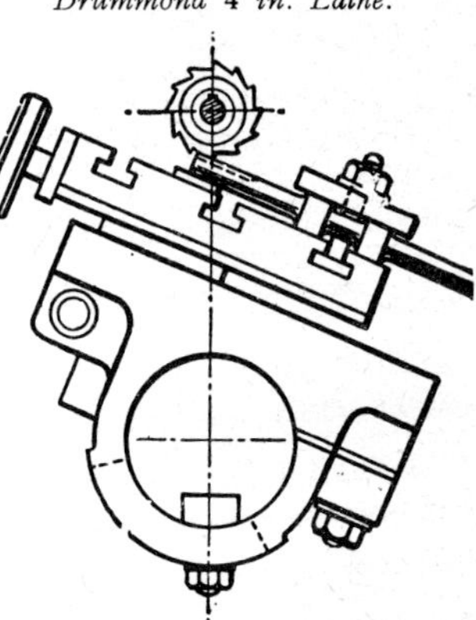

Milling a keyway in a shaft.

THERE are many machining jobs that are somewhat difficult to rig up—unless you have a Drummond Lathe, with the handy tee-slotted saddle and topslide, the feature of height adjustment by partial rotation of the saddle, the angle plate, and so on. This lathe makes work-setting easy. It is designed for boring, milling, and many other operations besides turning and screwcutting. It is very strong and rigid—the cylindrical bed is 3 ins. in diameter, heavy cast iron. Write for illustrated booklets (post free).

DRUMMOND BROS., LTD.

Rise Hill, GUILDFORD.

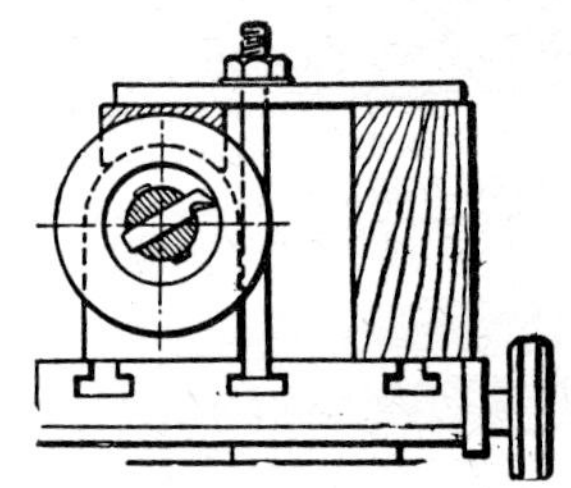

Boring a steam engine cylinder.

£4 18s. 6d.

3″ Centre Lathe

2-ft. gap-bed, swings 9″, complete with compound swivelling slide-rest

Foot Motor ... ... 35s. extra.

DUCK & HUTTON,
48, Percy Rd., N. FINCHLEY, N.,
Or call at STAND No. 45.

DRUMMOND

4in. LATHE USERS.

Bore your 12″ wheel and odd shape castings on our Eye Centre and Speed Reducer. Also Back Gear, Milling Attachments and Steadies.

3d. Stamps for Particulars.

MARLBORO ENGINEERING WORKS,
HORTUS ROAD, SOUTHALL.

ROUND BELTING.

"Torque" Belting and Fasteners for Lathes, etc., adjusted without cutting. Belting 5/16″ diam., 4d. per ft.; Fasteners, 9d. per set; ¼″ diam., 3d.; Fasteners, 8d., fitted. State exact length required. From 3/16″ to ½″ diam. Also "Dobby" Harness Cords, with Patent Couplings.

CHEELD BROS., *Engineers,* CLACTON, ESSEX

ESTABLISHED OVER 75 YEARS.

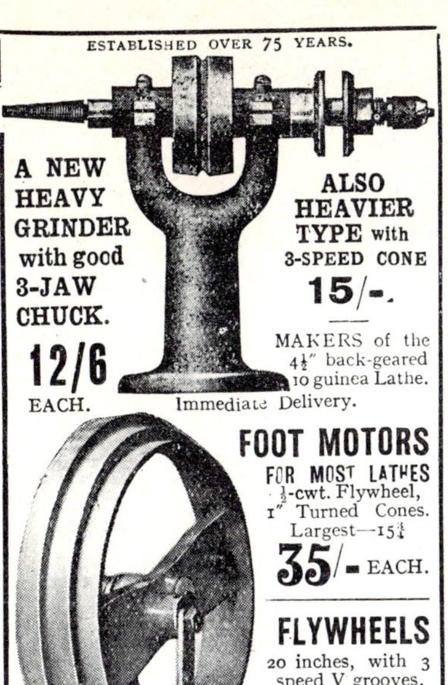

A NEW HEAVY GRINDER with good 3-JAW CHUCK.

12/6 EACH.

ALSO HEAVIER TYPE with 3-SPEED CONE 15/-.

MAKERS of the $4\frac{1}{2}$″ back-geared 10 guinea Lathe.

Immediate Delivery.

FOOT MOTORS

FOR MOST LATHES
½-cwt. Flywheel,
1″ Turned Cones.
Largest—15¼

35/- EACH.

FLYWHEELS

20 inches, with 3 speed V grooves,

20/- EACH.

Also LATHE STANDARDS and PEDALS.

EDWINSON GREEN & SONS, Ltd.,
CHELTENHAM, Eng.

Order Now.

Quick Delivery.

MAKERS:

EDWINSON GREEN & SONS, CHELTENHAM.
Or any Tool Dealer.

AN IDEAL TOOL

FOR
DRILLING,
GRINDING,
POLISHING,
SAWING.

3-speed Head and heavy 3-speed turned flywheel, 20 in. diameter. Height 40 ins. Substantial cast-iron pedestal and belt.

PRICE, with Adjustable Bar, T rest, Back-Centre, 3-jaw Chuck and belt,

75/-

SEE IT. *THE NEW*

EDWINSON GREEN
5-inch Back-geared, Hollow Spindle,
15 Guineas complete,
at STAND No. 6,
or Messrs. JONES TOOL CO.,
127, Old Street, LONDON, E.C.2, or at
87, High Street, CHELTENHAM.

CAN A REALLY HIGH CLASS 4″ B.G. S.C. LATHE BE MADE AND SOLD FOR £30 ?

Inspect the "ETA" Lathes on show at the M.E. Exhibition at Messrs. Buck & Ryan's Stand No. 16 and judge for yourself.

These Lathes are taken from ordinary stock, and finished exactly as supplied to the customer.

THE ETA TOOL CO.,

70a, Asylum Street,
LEICESTER.

SEE the New 3⅜″ and 4″ Screwcutting

SUPER-EXE LATHES

at the "M.E." EXHIBITION,

STAND No. 37

Also our 2½″ and 2″ LATHES.

"EXE LATHES EXCEL"

because of their accuracy and unique features.

IF UNABLE TO CALL, WRITE FOR LISTS TO:

The EXE ENGINEERING CO., LTD., EXETER.

AN

EXE 2½″ 3⅜″ or 4″ LATHE SIMPLIFIES SCREWCUTTING

CLOSE-UP VIEW OF HEADSTOCK OF 3⅜″ SUPER EXE LATHE

Note also divided pulley.

Hire purchase can usually be arranged.

This is usually a tedious operation, even to professional mechanics, and to amateurs is sometimes an absolute bugbear owing to the difficulty of picking up the thread in pitch. The special clutch on the mandrel of EXE Lathes, however, picks up automatically ALL threads (including uneven and fractional pitches) EVERY time in pitch, thus making screwcutting as easy as ordinary turning. This is uniqne among small lathes.

Bench Lathe from **£7 . 10 . 0**

(Carriage paid within 200 miles of Exeter)

The EXE ENGINEERING CO., LTD. EXETER.

EXE LATHES 4″ 3⅜″ 2½″

SOME OF THEIR UNIQUE FEATURES:

AUTOMATIC SCREW-CUTTING CLUTCH, which picks up all threads in pitch every time. Makes screw-cutting easy.

DIVISION PLATE on pulley. Most useful for marking off.

BACK-GEAR RATIO by large pulley, giving smoother running and making no noise.

BED, SADDLE AND SLIDES are precision surface ground.

EXE ENGINEERING CO. LD. EXETER.	**"M.E." EXHIBITION SAMPLE STAND.**

A FEW

of many Special Tools from

EXETER TOOLS AND MACHINERY, LTD.,

ALPHINGTON ROAD, **EXETER.**

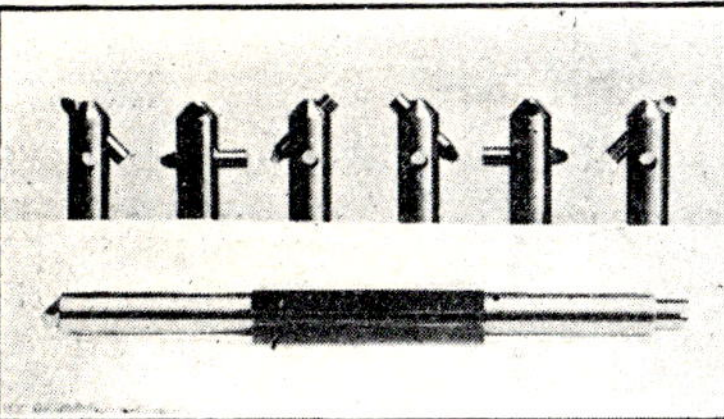

The "NuLok" TOOLHOLDER.

The most universal made. Note the numerous cutter positions. Sizes to suit all lathes.
WRITE FOR LIST.

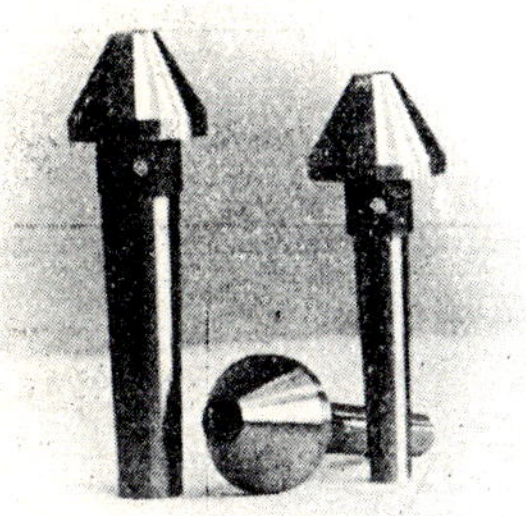

The "EXE" COMBINATION LATHE CENTRE.

Combining a large male centre for hollow work and a small female centre.

No. 1 MORSE SHANK, 5/6 each or 10/6 pair.
„ 2 „ „ 6/6 „ „ 12/6 „
Post free.

The "EXE" DRILL PADS.

Diam. 2", with Vee groove for drilling round rods, etc.

No. 1 MORSE SHANK, 6/9 post free.

The "Atlas" TURRET HEAD TOOL

(ALMOND PATTERN).

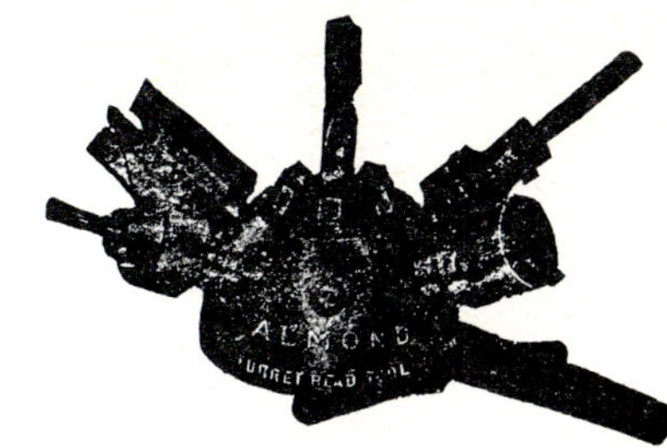

This attachment has a taper shank to fit the Tailstock spindle of an ordinary lathe, the turret revolves upon a large bearing surface on the body of the tool, and when locked cannot lift or spring during cutting. The locking device holds it in correct alignment, and the turret cannot be moved unless the lever is depressed.

SPECIFICATION. No. 1 size. 3½ in. diameter.
Socket Holes, ½ in. diameter, 1 in. deep. Weight 3½ lbs.
Fitted with Rough Shank .. .. **£1 11s. 6d.** nett.
„ No. 1 Morse Taper Shank **£1 13s. 0d.** „
„ No. 2 „ „ „ **£1 14s. 6d.** „

SPECIFICATION. No. 2 size. 5½ in. diameter.
Socket Holes, ⅞ in. diameter, 1½ in. deep. Weight 12 lbs.
Fitted with Rough Shank .. .. **£3 3s. 0d.** nett.
„ No. 2 Morse Taper Shank **£3 5s. 3d.** „
„ No. 3 „ „ „ **£3 6s. 9d.** „
Tools, Drills and Chucks, as shown in illustration, extra.

ARTHUR FIRTH,
Atlas Works, CLECKHEATON.

The *Febson* Test Indicator

Provisional Patent No. 2837/33.
BRITISH MADE.

This Indicator will register to 1/1000 part of an inch any variation in the surface level or diameter of an object. It is especially suitable for attachment to the pillar of a surface gauge, but may be fitted to any suitable form of vertical pillar for use on machines, etc.
The Tool is extremely strong in construction, reliable, sensitive and accurate. Made in nickel silver throughout, with hardened steel studs and pivots.

Price 9/- post free.

The Febson Manufacturing Co., 122, Kinveachy Gardens, Charlton, London, S.E.7.
or the principal tool dealers.

"DESOUTTER" Standardised DIE SETS.

Buy your PILLAR DIE BOLSTERS ready made—with ground faces.
PRICES CARRIAGE FORWARD.

The Die Set illustrated above is a No. 6 9"×7" Heavy Base, Light Top.
Write for further particulars.
MADE BY— 'Phone: Colindale 6346-7-8.

Size No.	Die Space.	Thickness of Punch Holder and Base. cast iron.		Co'plete Die Set, Light type. Net.			Extra for Heavy Punch, Holder or Base.	
	in.	light	h'vy	£	s.	d.	s.	d.
1	4×3	1"	1¼"	1	18	6	1	0
2	5×4	1¼"	1½"	2	5	0	1	3
4	7×5	1½"	2"	3	4	0	3	10
6	9×7	1¾"	2¼"	4	5	0	6	3
8	12×9	2"	2½"	6	12	6	10	8
10	16×12	2¼"	2¾"	8	3	0	16	0
4L	7×3	1¼"	1¾"	2	18	6	2	9
6L	9×4	1½"	2"	3	11	0	3	10
8L	12×5	1¾"	2½"	5	2	6	9	2
10L	16×6	2"	2¾"	6	7	6	14	0

DESOUTTER BROS., Ltd., The Hyde, HENDON, LONDON, N.W.9.

ARTHUR FIRTH, Toolmaker, Atlas Works, Cleckheaton

THE "ATLAS" BENCH SENSITIVE DRILLING MACHINE.

Made in 3 Sizes, as Illustration.

Specification.—These machines have been designed to meet the requirements of amateurs, etc., and are fitted with Yankee Pattern Rack and Lever Feed in all sizes. The machines have 2 speeds for flat belt and all have fast and loose pulleys. Each size has well-slotted circular table, and the base is also well slotted.

No. **0** size 6″ Drilling Machine; capacity to ⅜″ holes.

Complete	£2	18	6
Rough Castings and Drawings	0	13	0
Planing Table	0	2	3

No. **1** size 8″ Drilling Machine; capacity to ½″ holes.

Complete	£4	7	6
Rough Castings and Drawings	0	19	0
Planing Table	0	3	0

No. **2** size 10″ Drilling Machine; capacity to 9-16″ holes.

Complete	£5	10	0
Rough Castings and Drawings	1	7	6
Planing Table	0	3	6

SET OF MATERIAL TO CONSTRUCT INDEPENDENT 4-JAW CHUCKS.

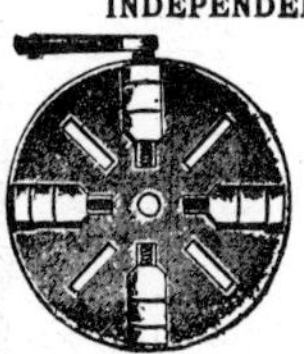

4″ dia. **4/-** per set consisting of strong cast-iron chuck-plate, 4 drop-forged steel jaws, 4 steel screw blanks, drop-forged box key, and full-size drawings.

4½″ „ **4/3** „
5″ „ **4/6** „
6″ „ **5/6** „
7″ „ **6/3** „
8″ „ **7/-** „

Heavier Pattern complete Sets, including Drawings.

7″ diameter, price	£0	7	6
8″ „ „	0	8	6
9″ „ „	0	9	6
10″ „ „	0	15	0
11″ „ „	0	17	0
12″ „ „	0	19	0
14″ „ „	1	3	0

All Sets have Drop-Forged Steel Jaws.

THE "ATLAS" VERTICAL SLIDE.

For 3½″ centre Lathes. Specially designed to fit the noted 3½″ centre Drummond Lathe.

Specification.—Table, size 5″ by 4″, with three machine-cut tee slots in same; slide is swivelling pattern, graduated and indexed for angle work; table is raised and lowered by square-thread screw, cut 10 threads per inch, and is fitted with American pattern ball handle. Workmanship of the best.

Price, complete	£1	8	6
„ Castings, rough	0	3	6
„ Planed Castings	0	10	6
„ Planed Castings, **including screw** and handle finished and graduating slide	0	18	6

SELF-ACTING HAND BENCH SHAPING MACHINES.

Made in 2 Sizes only.

No. 1.—8″ Stroke, 10″ Traverse. Rack and **Segment-driven** Table, 7″ square, well slotted.

Finished Machine	£6	0	0
Rough Castings and Drawings	1	3	6
Planed „ „	3	5	0

No. 2 Size (Extra Heavy)—8″ Stroke, 10″ **Traverse.** Table, 8″ square, well slotted.

Finished Machine	£7	0	0
Rough Castings and Drawings	1	15	0
Planed „ „	3	10	0

THE "ATLAS" SHAPING ATTACHMENT

For 3″ to 4″ centre Lathes.

No. 1 Size.

5″ Stroke, 8″ Traverse, Weight 40 lbs.

Strong Adjustable Slotted Table, 5″ square; Ram is graduated and indexed for angle work, and is driven by drop-forged steel connecting-rod. Tool box takes ⅜″ square tools, and has relief for tool.

Price, complete	£3	15	0
Complete Set of Rough Castings and Drawings	0	15	6
Complete Set of Planed Castings and Drawings	1	16	0

This attachment has been specially designed to suit the noted 3½″ Drummond Screw-cutting Lathe. (Two other sizes. See list.)

THE "ATLAS" BENCH MILLING MACHINE.

Specification: Table is 4″ wide, 12″ long, and has 2 machine-cut T slots in same, and has a longitude traverse of 8″ by 4″ the other way, and will fall 6″ below centre to top of Table. Spindle is bored No. 2 Morse Taper. Cone has 3 speeds for 1½ flat belt.

Complete with overhanging **arm as shown.** Weight, 150 lbs.

Price of Finished Machine	£10	10	0
Complete Set Rough Castings & Drawings	1	17	6
„ „ Planed „ „ „	3	15	0

SPECIALITY:

3″ Centre Lathe Sets of Parts.

Complete Sets of 3″ Back-geared Lathe Headstocks, Rough Castings, cast gears.
Price .. .. **6/3** per set.

Ditto, with heads bored and planed, and all gear-wheels turned and machine-cut teeth.
Price .. .. **20/-** per set.

Ditto, but 3″ centre heads, accurately **finished. 45/-** per pair.

Suitable 30″ Planed Gap Lathe Bed (no gap-piece fitted), 10/- each.

Ditto, with gap-piece fitted, **14/-** each.

For saddle and nut box castings, etc., and all parts to convert into Screw-cutting Lathe, see page 5 of List.

IMPROVED HAND PLANING MACHINES, With Self-acting Feed.

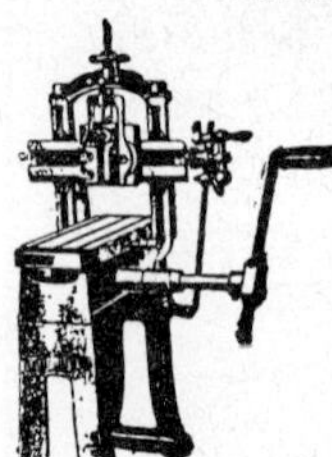

No. 0 size, for bench, planes 16″ long, 6″ wide by 4½″ high. Rough Castings, **25/-** set. Planed Castings, **63/-** set. Finished Machine, **£7 5s.**

No. 1 size, planes 18 by 7 by 5. Rough Castings, **30/-**; Planed Castings, **68/-**; Finished Machine, **£9 10s.**

No. 2 size, planes 20 by 7 by 5. Rough Castings, **35/-**; Planed Castings, **72/6**; Finished Machine, **£10 5s.**

No. 3 size, planes 24 by 10 by 7. Rough Castings, **58/6**; Planed Castings, **£5 17s. 6d.**; Finished Machine, **£16 10s.**

Drawings for any size, **3/6** set extra.

Standards instead of Bench Legs, any size of machine, rough or planed or finished, **12/-** extra.

SPECIAL HAND PLANING MACHINE

Planes 14″ long, 6″ wide, and 4½″ high. Price of complete finished machine, **£5.**

IMPROVED BENCH TWIST DRILL GRINDER.

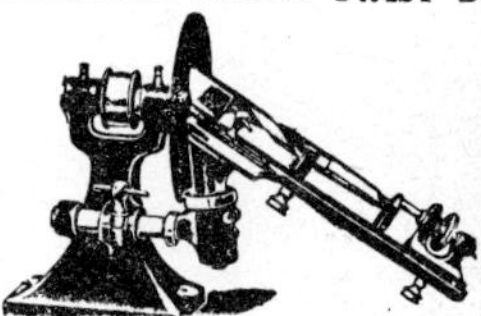

Grinds Drills 0 to 1″ diam. Complete set of Rough Castings, including steel for shaft, 2 oil-hole covers, and full-size blue prints, **14/6** set.

SELF-ACTING POWER SHAPING MACHINE.

This machine has 10″ stroke and 12″ traverse, all teeth are machine-cut from solid, and all **T-slots** are machine-cut from solid; weight, 6 cwts.

Price Rough Castings, **£6 15/-**
Price Planed Castings, **£12.**
Finished machine, **£21.**

IMPROVED BENCH TOOL GRINDER.

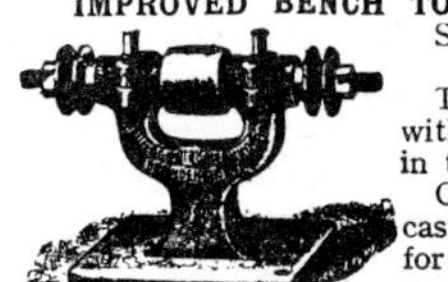

Suitable for wheels 8″ by 1″.

The machine is fitted with two rests not shown in the illustration.

Complete set of rough castings, including steel for shaft **&** two oil-hole covers. Price, 7/6 set.

Price finished machine, with **two 8″** dia. wheels, **41/-**

THE "ATLAS" DOUBLE-GEARED DRILLING SPINDLE.

Body hardened steel, 1″ square, 5″ long, machine-cut steel gears; nose ⅜″ Whit. thread; specially designed for **Drummond 3½″** lathe.

Price **£1 7 0**

Our 1912 Catalogue, 100 pages, of Lathes, American Chucks, Bench Grinders, Bench Drilling and Milling Machines, Face, Angle, and Surface Plates, Twist Drill, Lathe Beds, Slide-rests, Change Wheels, Slide-rest Tools, Lathe Carriers, Lathe Heads, Slot Drilling Slides, Wheel Cutting Attachments, Tool Holders, End Mills, Steel Shafting, Screwing Stocks, Rough and Machined Castings. Post free, **1s.**, which is allowed off first order over **7/6** value.

SELF-ACTING HAND BENCH SHAPING MACHINE.

5" stroke, 12" traverse, rack and segment driven table, 8" square, complete, with vice.

Finished Machines, with Vice	£7 5 0
Complete Set of Rough Castings, including Drawings	35/-
Planed Castings and Drawings	70/-

IMPROVED HAND PLANING MACHINES, With Self-acting Feed.

No. 0 size, for bench, planes 16" long, 6" wide by 4½" high Rough Castings **25/-** set. Planed Castings, **63/-** set. Finished machine, **£7 5s.**

No. 1 size, planes 18 by 7 by 5. Rough Castings, **30/-**. Planed Castings, **68/-**. Finished Machine, **£9 10s.**

No. 2 size, planes 20 by 7 by 5. Rough Castings, **35/-**. Planed Castings. **72/-**. Finished Machine, **£10 5s.**

No. 3 size, planes 24 by 10 by 7. Rough Castings, **58/6.** Planed Castings, **£5 15s.** Finished Machine, **£16.**

Drawings for any size **3/-** set extra.

Standards instead of Bench Legs, any size of machine, rough or planed or finished, **7/6** extra.

WHITON'S CHAMPION INDEPENDENT CHUCKS WITH REVERSIBLE JAWS.

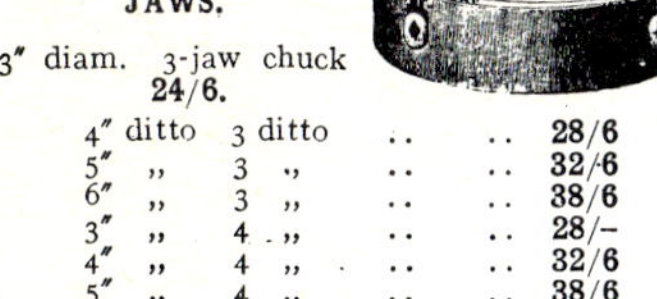

3" diam.	3-jaw chuck	**24/6.**
4" ditto	3 ditto	28/6
5" ,,	3 ,, ...	32/6
6" ,,	3 ,, ...	38/6
3" ,,	4 ,, ...	28/-
4" ,,	4 ,, ...	32/6
5" ,,	4 ,, ...	38/6
6" ,,	4 ,, ...	43/6

IMPROVED BENCH TWIST DRILL GRINDER.

Grinds Drills 0 to 1" diam.

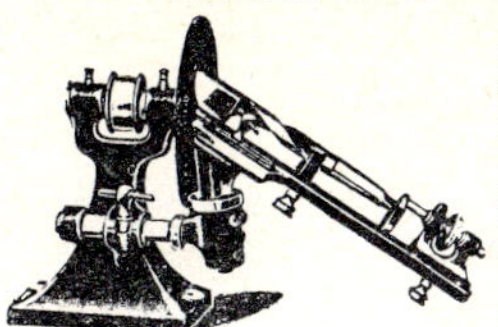

Complete set of Rough Castings, including steel for shaft, 2 oil hole covers, and full size blue prints, **14/6** set.

Boxed and free on rail.

ACCURATELY MADE COMPOUND SLIDE-RESTS

Suit centre Lathe.	Rough Castings.	Planed Castings.	Finished Rest.
2½"	**2/3**	**9/6**	**35/-**
3"	**3/-**	**10/6**	**40/-**
3½"	**3/6**	**11/6**	**45/-**
4"	**5/3**	**14/6**	**54/-**

Cutting swivel slide out for **T** bolts, any size of planed castings, **2/3** extra.

VERTICAL SLIDES FOR MILLING, Etc., IN THE LATHE.

We will include a suitable horn handle with each set ordered through this advertisement.

Suitable for	Rough Castings. Per Set.	Planed ditto.	Finished Slide.
2½ and 3" centre lathe	**3/6**	**12/-**	**36/-**
Suit 3½" and 4" lathe	**4/9**	**14/6**	**38/6**
Suit 4½" and 5" lathe	**5/6**	**16/6**	**45/-**
Suit 6" Lathe ..	**10/6**	**23/-**	**56/-**

SLIDE-REST TOOLS.

Made of best cast steel, assorted shapes, and ready for use.

¼" square, **3/9** dozen; 5/16", **4/3**; ⅜", **5/-**; 7/16", **6/3**; ½", **7/6.**

Self-hardening high-speed tools. Set of 6 tools at prices of dozens. For tools for shaping and planing machines and larger sizes, see list.

CUSHMAN'S AMATEUR LEVER SELF-CENTREING CHUCKS

3" diam. with 2 sets of jaws .. **22/-**

4" ditto, ditto .. **25/-**

5" ditto, ditto .. **27/6**

Prices with 1 set of jaws—3", **17/6**; 4", **20/-**; 5", **23/-**.

SET OF MATERIAL TO CONSTRUCT INDEPENDENT 4-JAW CHUCKS.

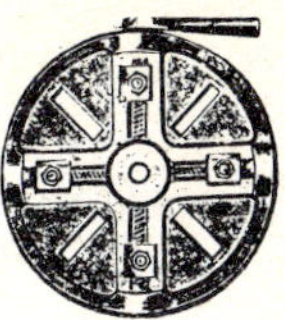

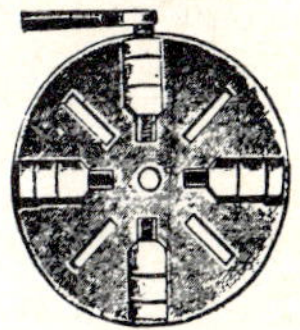

Consisting of strong cast-iron chuck-plate, 4-drop forged steel jaws, 4 steel screw blanks, drop forged box key, and full-size drawings.

5" dia. **5/6** per set
6" ,, **6/3** ,,
7" ,, **7/-** ,,
8" ,, **8/-** ,,

NOTE.—We have had the dies specially made for the jaws and box keys. For larger sizes, see list.

LATHE FACEPLATE DOGS.

A Set of Four of these convert an ordinary Faceplate into a handy chuck.

To Fit Slot. Width in ins.	Suitable for diam. of Faceplate.	Price, Set of 4 Finished Dogs.
5/16"	5"	**7/-**
⅜"	6"	**7/6**
7/16"	7"	**8/-**
½"	8"	**8/9**

CAST-IRON V-BLOCKS.

Accurately planed in pairs.

1" long by	⅝" square ..	**2/-** pair.
1½" ,,	¾" ,, ..	**2/6** ,,
2" ,,	1" ,, ..	**3/-** ,,
2½" ,,	1¼" ,, ..	**3/3** ,,
3" ,,	1½" ,, ..	**3/6** ,,

SPECIAL HIGH-SPEED SELF-HARDENING CUTTING-OFF STEEL.

Flat section 1/16"×¼", **8d.** ft. 3/32"×½", **10d.** ft.

For Square and Round sizes, see List.

LATHE FACE PLATES CLAMPS.

Set of 4 Clamps 5/16" size, and set of 10 suitable bolts suitable for 3 and 3½" centre lathes, **2/6** set. ⅜" size for 4" lathes, **3/-** set.

BEST BRIGHT DRAWN STEEL RODS (Round).

Twelve	one-foot lengths	⅛" to ½" ..	**1/4** bundle.
,,	two ,, ,,	,, ..	**2/6** ,,
,,	three ,, ,,	,, ..	**3/8** ,,
Eight	one ,, ,,	¼" to ¾" ..	**2/-** ,,
,,	two ,, ,,	,, ..	**3/9** ,,
,,	three ,, ,,	,, ..	**5/6** ,,
Ten	one ,, ,,	¼" to 1" ..	**3/-** ,,
,,	two ,, ,,	,, ..	**5/6** ,,
,,	three ,, ,,	,, ..	**8/-** ,,
Twenty-five	one-foot lengths,	¼" to 1½" ..	**9/-** ,,
,,	,, two ,,	,, ..	**17/6** ,,
,,	,, three ,,	,, ..	**25/-** ,,

Thirty one-foot lengths, assorted shapes, flats, squares, hexagons and rounds .. .. .. **4/3** ,,

COMPLETE SETS OF BACK-GEARED LATHE HEADSTOCK ROUGH CASTINGS.

	Ordinary Heads	Screw-cutting Heads		Ordinary Heads	Screw-cutting Heads
2½" centre ..	**5/3**	—	4½" centre ..	**18/6**	**21/-**
3" ,, ..	**7/-**	**8/6**	5" ,, ..	**23/-**	**28/-**
3½" ,, ..	**7/9**	**9/6**	6" ,, ..	**30/-**	**36/-**
4" ,, ..	**12/-**	**14/6**			

THE "ATLAS" SHAPING ATTACHMENT,

FOR 3" TO 4" CENTRE LATHES. 5" Stroke. 8" Traverse. Weight 40 lbs.

Strong Adjustable Slotted Table, 5" square; Ram is graduated and indexed for angle work, and is driven by drop forged steel connecting-rod. Tool box takes ⅜" square tools, and has relief for tool.

Price Complete	**£3 15 0**
Complete Set of Rough Castings and Drawings	**£0 15 6**
Complete Set of Planed Castings and Drawings	**£1 16 0**

This attachment has been specially designed to suit the noted 3½" Drummond Screw-cutting Lathe.

NEW SPECIALITIES.

BENCH MILLING MACHINE CASTINGS, with overhanging arm, table 4" wide, 12" long, 3-speed cone 1½" flat belt.

Rough Castings and Drawings .. .. **35s.**
Planed ,, ,, ,, ... **£4 10s.**

8" SENSITIVE DRILLING MACHINES, Yankee pattern, rack-feed, spindle bored. No. 1 Morse taper. Capacity, ½" holes.

Castings and Drawings .. .. **15s.**
Finished Machine .. .. **£4 7s. 6d.**

10" SENSITIVE DRILLING MACHINES. Same design as 8" above. Capacity, 9/16" holes.

Castings and Drawings .. .. **25s.**
Finished Machine .. .. **£6.**

Send 6d. for my New 86-page 1909 Catalogue of Engineers' and Amateurs' Tools. Allowed off first order of 7/6 and upwards. Colonial and Foreign Readers sending 8d. will receive Catalogue by Registered post.

ARTHUR FIRTH, Toolmaker, Cleckheaton, Yorkshire, England.

THE DART EMERY WHEEL

TWIST DRILL GRINDER.

The "Dart" Twist Drill Grinder provides an inexpensive appliance for accurately grinding Twist Drills, from 1-16th in. to ½ in. As it frequently happens trouble from breakage is frequently experienced with these small drills, which is often caused by their not being correctly ground, this machine will soon repay its cost in any Works where these small drills are used.

Write for Catalogue No. 176.

Sample Machine sent for trial on approval, carriage paid to any point in the U.K. upon receipt of £3 12s. 6d.

Glasgow Office: 9, HOWARD ST.

THE FAIRBANKS COMPANY,
78-80, CITY ROAD, LONDON, E.C.

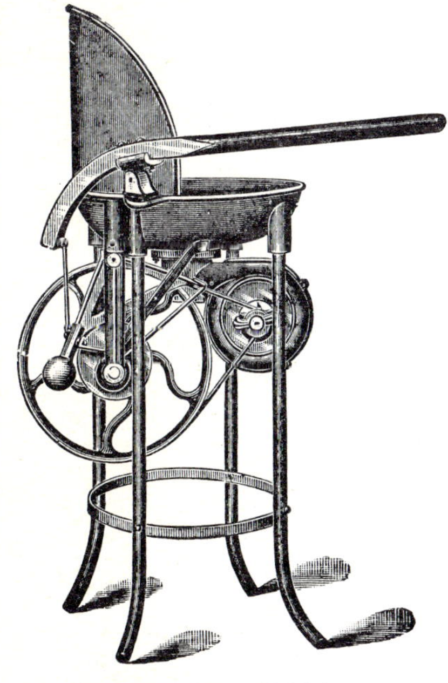

No. 21 EUREKA.
Price 56/3.

Potts' Portable Forges.

SIMPLE, DURABLE, CONVENIENT.

THIS Forge is well adapted for small workshops. It takes but little room, the floor space required is 2 ft. × 2 ft., height 33 ins. It can easily be moved about. It will bring a one inch rod to a welding heat in six minutes. It is far more convenient than a bellows forge, the blast being more powerful, with less labour. It enables the Amateur to practice the most interesting and fascinating work in Engineering practice, and saves a great amount of filing and grinding, and to use wrought iron and steel where he would otherwise be tempted to use castings.

Sent carriage paid upon receipt of price to any point in the U.K.

WRITE FOR CATALOGUE.

THE FAIRBANKS CO.,
78-80, CITY ROAD, LONDON, E.C.
Glasgow Office: 9, HOWARD ST.

HIGH-CLASS ACCURATE

PLANING MACHINES.

Capacity 14″×6″×5″. Weight 100 lbs. net.
An efficient Planer of rigid construction for Model Making and Shop use. All angles 50° either side of Zero can be machined.

Price £6 Complete. Vice 12/6 extra.

Packing case charged 4s.; allowed in full if returned carriage paid. Full particulars Post Free, 1d. Stamp.

12/6 BORING CARRIAGES, 15/-

For Boring Cylinders, Dynamos, Bearings, &c., on an ordinary Lathe. Send sketch of Lathe bed with dimensions, and 1d. stamp for photo.

C. W. FRANKLIN, SMALL TOOL MAKER,
CLEVELAND ROAD, UXBRIDGE.

2-SPEED HAND PLANERS

Price **£13 10** complete.

With adjustable automatic friction feed.

Machine cut gears and rack.

Capacity: 18 in. × 8in. × 6 in.

C. W. FRANKLIN, Small Tool Maker,
CLEVELAND ROAD, UXBRIDGE.

LATHES, TOOLS,
FILES, GEARS,
MODEL ENGINE
SUPPLIES

described in Catalogue 1842x, mailed for 2d. stamp

The FRASSE CO., 38, Cortlandt St., New York, U.S.A.

LATHES.

6 ins. swing, 12 ins. between centres, six Turning Tools and Drill Chuck. Without Stand, **14 dollars**; Mounted on Stand, **19 dollars.**

Catalogue 1842x describes **FILES, GEARS, PINIONS, MODELS,** mailed for 2d. or 2 cents. U.S.

THE FRASSE COMPANY, 38, Cortlandt Street, New York, U.S.A.

Postage to America, 2½d.

DON'T WASTE TIME

changing tools:

Fit a Square Turret to your Lathe
and increase the amount of your spare time by 100%. Automatically registers the operating tool, enables the operator to turn, bore, face, and cut off without touching a spanner.

For **4¼″ Britannia** ...	**£2 : 0 : 0**
,, **3½″ Drummond** ...	**35/-**
,, **4″ ,,**	**25/-**
,, **2½″ Exe**	**17/6**

Also made promptly to fit all other makes.

Goodman's Engineering Works,
FROGNAL AVENUE, WEALDSTONE, Middx.
Send us your inquiries for general machining work.

WANTED

Machining Work of all descriptions, Turning, Milling, Precision Grinding, etc. Also Production Work.

GOODMAN'S ENG. WORKS,
Manufacturers of Small Tools and Motor Productions,
Frognal Avenue, WEALDSTONE, Middx.

DON'T WASTE TIME

changing tools:

Fit a Square Turret to your Lathe
and increase the amount of your spare time by 100%. Automatically registers the operating tool, enables the operator to turn, bore, face, and cut off without touching a spanner.

For **4¼″ Britannia** ...	**£2 : 0 : 0**
,, **3½″ Drummond** ...	**35/-**
,, **4″ do.**	**25/-**
,, **2½″ Exe**	**17/6**

Also made promptly to fit all other makes.

Another Real Time Saver.

Geared drilling and milling spindles, hand operated (no overhead required) as well as power. A tool capable of milling keyways, link motion parts, drilling etc. An extremely useful accessory in your workshop. Complete with chuck. Each **20/-**

Particulars of our new treadle sensitive Drilling Machine sent on request.

We specialize in all tools and accessories for the amateur engineer.

Goodman's Eng. Works,
FROGNAL AVENUE, WEALDSTONE, Middx.

VISITORS TO OUR STAND

At "**The Model Engineer**" **Exhibition** were astounded at the marvellous value we offer, and asked us how it is possible to do it. We can only repeat, that it is making a quantity, on the best tools, by modern methods, and having the whole of the processes (from the Pig Iron to the finished tool), under our own care, that we are able to do it. All middle profits are cut out and the customer gets the benefit in increased value. A few of our lines are here mentioned.

OUR POPULAR INDUSTRY LATHE

is a real wonder, and is still in front. We claim this to be the finest value ever offered. It is made in two sizes, and the price is extremely reasonable. **Specification:** 3-in. centre on 2 ft. planed bed; headstock is a strong casting fitted with phosphor-bronze bearings, which are adjustable. The spindle is of large size, accurately fitted, and has 3-speed turned and grooved cone pulley. The thrust is taken by our improved method, and is adjustable. The tailstock is of the overhung pattern, fitted with steel sliding barrel, operated by a machine-cut steel screw and polished hand wheel. Split grip lock is provided. Each lathe is complete with face-plate, hand-rest, and two tees. Prices are as follows: 3-in. centre, 2 ft. bed, 17s. 6d.; if with hollow spindle to pass $\frac{3}{8}$-in. bars, 20s.; $3\frac{1}{2}$ ins. centre, 2 ft. 6 ins. bed, with hollow spindle, 30s.; treadle motion supplied complete, with polished tool tray, for 3-in. lathe £1 extra; for $3\frac{1}{2}$-in. lathe 25s. extra. Our Jenwyn compound swivelling slide rests can be fitted to the above lathes, 3-in. centre, 17s. 6d.; $3\frac{1}{2}$-in. centre, 21s.; 3-in. lever scroll chuck with two sets of jaws, fitted complete, 22s. 6d. extra. Our slide-rests have created quite a stir in the tool world. They have long wide slides, square thread screws, cut from steel bars, and loose strips to compensate for wear. Sent willingly on approval. Money back if not satisfied.

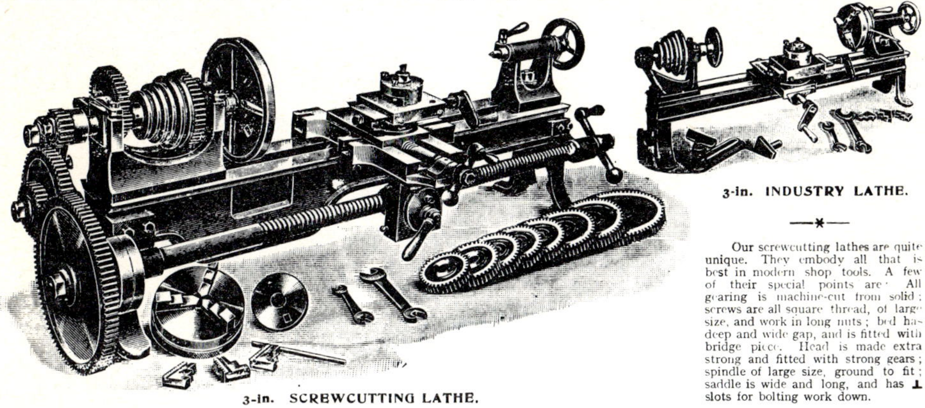

3-in. INDUSTRY LATHE.

—*—

Our screwcutting lathes are quite unique. They embody all that is best in modern shop tools. A few of their special points are: All gearing is machine-cut from solid; screws are all square thread, of large size, and work in long nuts; bed has deep and wide gap, and is fitted with bridge piece. Head is made extra strong and fitted with strong gears; spindle of large size, ground to fit; saddle is wide and long, and has ⊥ slots for bolting work down.

3-in. SCREWCUTTING LATHE.

Specification: 3-in. centre, back geared, slide and screwcutting lathe, on 2 ft. 6 ins. heavy gap bed, fitted with bridge piece. All surfaces are accurately machined, and hand-scraped to surface plate. Gearing machine cut from solid blanks. A really high-class tool at a low price. Prices: 3-in. centre on 2 ft. 6 ins. bed, bench lathe, £7 17s. 6d. Treadle lathe, £9 17s. 6d. Can be had on 3-ft. bed for 15s. extra. Also made $3\frac{1}{2}$-in. centres on 3 ft. 6 ins. heavy gap bed, as specification above. Bench lathe, £10 5s. Treadle lathe, £12 15s. Each lathe is complete with set of spanners, driver, and faceplate, and set of change wheels. We can fit a 4-in. lever scroll chuck, with two sets of jaws for 27s. 6d. extra. The 3-in. lathe, on 2 ft. 6 ins. bed, admits 18 ins. between centres, and 11 ins. diameter in the gap. The larger lathes admit larger in proportion. **Our New Catalogue is now ready,** forty-eight pages of interesting matter nicely illustrated. **Post free, 3d. stamp.**

GOODWINS', ENGINEERS, LEEK, STAFFS.

All the World Wondered—

Yes, wondered how it was possible to give such astounding value for money. 'Tis simple. We cut down expense to the lowest point. Each man we employ is an expert at his work. We have none but Modern High Speed Tools, and making these LATHES, in large quantities, **WE CAN GIVE YOU A BENEFIT,** if you will give us your orders. We are daily receiving Unsolicited Testimonials from purchasers of these Lathes, and they have obtained the highest opinion of professionals, as well as amateurs. We have recently received a number of letters asking if we still made the Lathe at 16/6. In answer, we may say, we do. In future **DON'T WAIT** for our advt. to appear, but SEND YOUR ORDER right along. We have always a batch in hand in the works, and at the longest shall only keep you waiting a few days. We have a number now ready, so if you want one, send at once, YOU WILL BE WELL SATISFIED.

POINTS TO REMEMBER.

Bed accurately Planed. **Turned and Polished IRON, Cone Pulley.**
All Parts accurately Fitted. **Centres of Tool Steel, and Interchangeable.**
Up-to-Date Design, Large Wearing Parts.

ONLY 16/6.

THE PINNACLE OF VALUE.

ONLY 16/6.

GOODWIN, Engineer, LEEK.

MECHANICS $2\frac{1}{2}$ in. Cen. LATHE.

We have now REVISED PRICES for the component parts for this lathe. A descriptive leaflet will be sent post free on request.

Owing to the numerous enquiries for a Slide Rest, we have now perfected one for this lathe, price

45/- post free.

This SLIDE REST is fitted with V guides and provision for taking up wear. Cross slide can be turned to any angle, the whole is adjustable for height and can be swung round to any angle without disturbing the adjustment for sliding.

E. GRAY & SON, LTD.,
18-20, Clerkenwell Road, LONDON, E.C.1.
ESTABLISHED 1822.

E. GRAY & SON, Ltd.

The most Useful Present for all Hobbies.

UNIVERSAL LATHE *(as illustrated.)*

comprising Headstock. Tailstock. DRILL CHUCK. 5-in. Faceplate. Long Tee Rest. 5-in. Emery Wheel. 3-speed Cone Pulley. 15-in. Bar Bed.

PRICE **21/6** Part postage 9d.

NOTE.—Accessories supplied with our Universal Lathe as previously advertised by us will interchange with the above lathe. Full price list of accessories obtainable sent with each lathe. We have only a limited number and this offer cannot be repeated.

Speed Boat Petrol Engine, The 'GRAYSON.'

1¼-in. bore. 1¼-in. stroke. 25 c.c.

This engine is designed on the existing models produced by Mr. F. N. Sharpe. Recent successes with boats fitted with these engines have been recorded. Speeds attained 27 and 26 m.p.h. This engine is fitted with carburettor and wiper contact. Bosch sparking plug.

PRICE **£6 : 10 : 0** SET OF CASTINGS (with Blue Print) **17/6** Postage 1/3.

We can supply at small extra cost any parts machined.

A brief list of various prices will be sent free on application.

18-20, Clerkenwell Road, London, E.C.1.

ESTABLISHED 1822.

My NEW SHAPING MACHINE.

6-in. Stroke (Adjustable). 8-in. Traverse (Automatic).

ADJUSTABLE TABLE. **Price £12 12s.**

EDWARD HINES, High Class Lathe Manufacturer, **NORWICH.**

Instal the New 1 h.-p. "Hartop" Industrial Engine in your Garden Workshop.

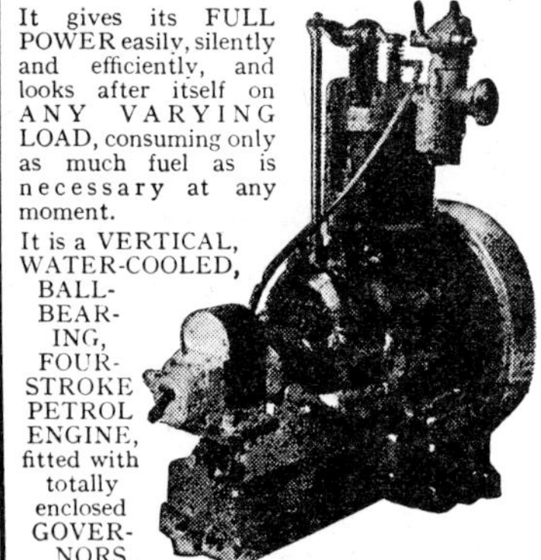

It gives its FULL POWER easily, silently and efficiently, and looks after itself on ANY VARYING LOAD, consuming only as much fuel as is necessary at any moment.

It is a VERTICAL, WATER-COOLED, BALL-BEARING, FOUR-STROKE PETROL ENGINE, fitted with totally enclosed GOVERNORS, carburettor, box bed, sparking-plug, silencer, handle and magneto-coupling, **at the amazingly low price of £8 18s. 6d.**

It is fitted with an AUTOMATIC LUBRICATION SYSTEM, together with MANY OTHER ADVANTAGES, and is fully described in our NEW ENGINE LISTS, which are POST FREE.

FRANK HARTOP & SONS, *Engineers,*
Miller Road, Bedford.
Telephone: Bedford 2977.

A REAL TOOL, NOT A TOY

Accurate Bench Lathe, 20" Bed, 2¼" Centres, Hollow Mandrel, Compound Sliderest, Hand Rest, 6 Collet Chucks taking up to 5/16", Bell Chuck, Driving Plate, Wood Turning ditto. Exquisitely finished. **Price 5 Guineas** Carriage Paid. Cash returned if not entirely satisfied.

S. HARRIS, Lathe & Engine Factor,
51, Upton Lane, Forest Gate, LONDON, E.7

PORTASS

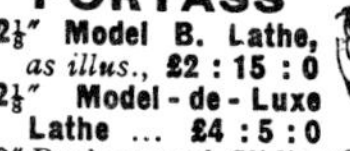

2½" Model B. Lathe, *as illus.,* **£2 : 15 : 0**
2½" Model-de-Luxe Lathe ... £4 : 5 : 0
3" Back-geared, Sliding, Self-acting Screwcutting, Lathe, with iron stand, 3-speed wheel, chip tray ... ... **£12 : 12 : 0**
Write first instance: **HEELEY MOTOR and MANUFACTURING Co., Broadfield Rd., Sheffield.**
Deferred Terms through: **J. G. GRAVES, Sheffield.**

SLOTTING MACHINE

6" geared auto traverse to all motions, modern high-class machine, together with the professionally-made patterns and three sets of castings part-machined. A sound proposition to a firm on the look-out for a good speciality.

H. HISTON, 121a, Constitution Hill, BIRMINGHAM.

JUST THINK IT OVER.
THE BEST

Is none too good. It may (and probably does) cost more money to buy than something of an inferior grade. But in the long run the **BEST IS THE CHEAPEST.**

This applies to goods of any description, but to nothing does it apply more forcibly than the **NEW HOLMES FIVE POUND LATHE.**

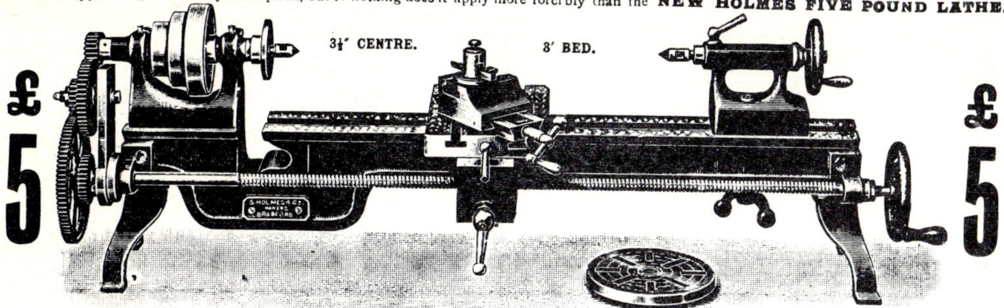

Self-acting, Sliding, Boring, and Screw-cutting Lathe, 3½″ Centres, 3′ Gap Bed, Depth of Gap 3″, size of Boring Carriage 8″ by 6″, with full set of change wheels; can be fitted with cone pulley ofr round belt if desired; complete as Treadle Lathe **£2** extra.

Every manufacturer advertises his products as the best. This is only natural. But when several concerns manufacture a similar article (**a Five Pound Lathe,** for example), it is obvious that they cannot all turn out the best machine on the market.

There can only be one best. The question is, how to pick it out. Well, we are ready to help you.

We refer you to our long standing as the pioneers of the cheap lathe. We refer you to the thousands of **unsolicited testimonials** we have received. We remind you that every lathe is thoroughly tested. We remind you that we employ only experienced mechanics. Made in the shops where tens of thousands of lathes have been made. Made by the latest up-to-date machinery. We send every lathe on approval, and if found **not up** to expectations, we will have the machine returned to us and pay you for your trouble.

There is not another **Five Pound Lathe** to come up to the New Holmes for accuracy, strength, and reliability.

ONCE AGAIN WE ASK YOU TO THINK IT OVER. Can there be a more honest way of dealing? Just sit down and write

HOLMES & CO., The Lathe Firm, BRADFORD,

Enclosing Stamp for Particulars, or Four Stamps for Illustrated Katalog.

"All the World" Slack. HOLMES & Co. Busy.

THE REASONS WHY, THESE:

FIVE POUNDS. FIVE POUNDS.

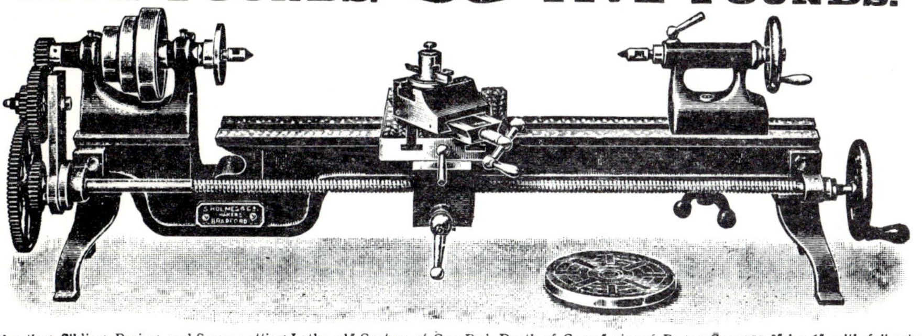

Self-acting, Sliding, Boring, and Screw cutting Lathe, 3½″ Centres, 3′ Gap Bed, Depth of Gap 3″, size of Boring Carriage 8″ by 6″, with full set of change wheels; can be fitted with cone pulley for Round belt if desired; complete as Treadle Lathe **£2** extra.

KATALOG 4 STAMPS.

TWIST DRILL GRINDER.
UP TO 1 INCH,

55/-

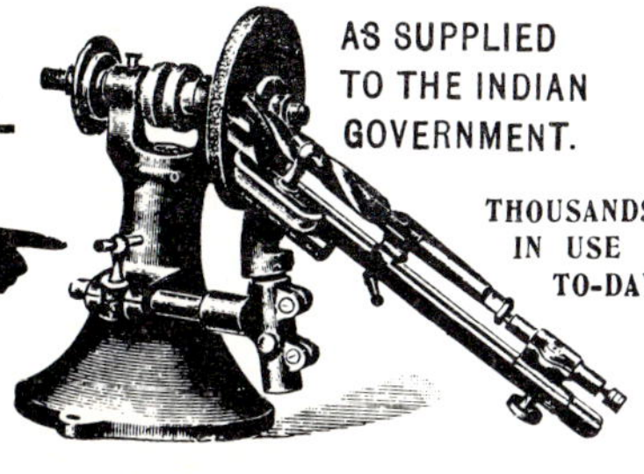

AS SUPPLIED TO THE INDIAN GOVERNMENT.

THOUSANDS IN USE TO-DAY.

HOLMES & Co.,

THE LATHE FIRM, BRADFORD.

FOOT POWER DRILLING MACHINE.

CAN BE SUPPLIED WITH FLAT BELT.

£6 10s.

AND NO WONDER.

It Stands on its Standards as Rigid as a Rock.

THE "NUMBER EIGHT" SLIDING, BORING AND SCREW-CUTTING LATHE FOLLOWS THE OLD LINES.

BUT the old lines have been tested, tried, and found true. Never mind about the new fangled ideas which are supposed to revolutionise all existing things. This Lathe of ours has stood the test, and every one sold has called forth words of praise. Would this be the case if it was not accurate? Talk about stability! It **stands on its standards as rigid as a rock.** Notice the whole thing! Look at the design! Specially thought out—not the result of a passing whim, a desire to be different from others. The outcome of years of thought. 3 ins. centres, 3 ft. heavy gap bed, gunmetal bushes, all wearing parts adjustable, MACHINE-CUT GEAR. The treadle, in the actual thing, extends from standard to standard. The crankshaft runs on hardened centres, and the whole **Lathe Stands Out** as the **Acme of Construction** at the price. Just as illustrated, **£10.** Fitted with nicely designed boring carriage, **15/-** extra. If you care, you can have it as an Automatic Sliding Lathe for **£8 15s.** Supplied with gear wheels, as shown on leading screw. If you do really want a first-class lathe at a price which is out of all proportion to its value, give us a trial. We guarantee satisfaction. Send for KATOLOG, Four Stamps. Have you seen the "LITTLE RED BOOK" of "GEM SPECIALITIES"? Better enclose Two Stamps for this.

S. HOLMES & Co., Engineers, BRADFORD.

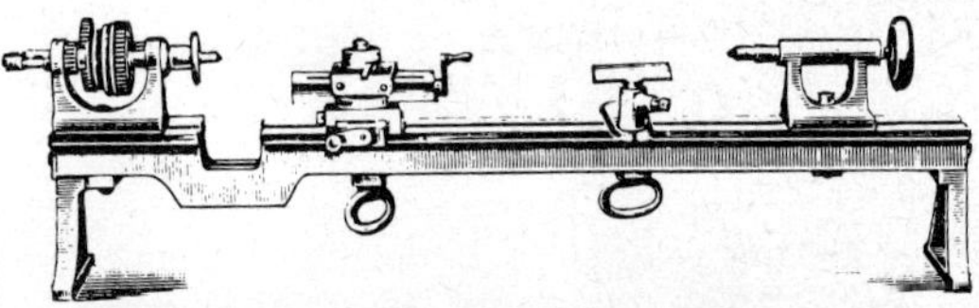

THE LITTLE SAMSON.

Now then, here is another little champion Lathe for you Model Makers. Not a toy, but a strong well-made lathe, two and a half inch centres, thirty inch cast iron bed. It is fitted with conical necks, the gear wheels are

Gun Metal,

and we guarantee this lathe to do accurate work. We have sold twenty-four this last fourteen days. The price is only **£2 10s.**, and its equal cannot be found. We supply it with a compound slide-rest for **25/-** extra. We can deliver this lathe in four days from order.

To Sing its Praises is to Gild Gold.

If you are in want of a Foot Motor, we are the firm to fit you up. You no doubt saw our block of one in the March 1st issue of the *M.E.*, price **25/-**.

17s. 6d.

also buys our now famous New Century Lathe. We surmise that by this time you are well aware that this lathe is a gift at the price. You had better send four stamps for our Katalog; you will receive it per return.

S. HOLMES & CO.,
Engineers and Tool Makers, **BRADFORD**

ONE! TWO!! THREE!!! OFF.
ANOTHER ROUND.

17/6

VOTE FOR VALUE.

Well, how goes it with your work? Shall we be hearing from you this year? If you have a want, and money to supply that want, and that want is a Lathe, we want you. The **No. 1 Special,** which you see above, is the neatest, handiest little tool on the market. We have sold hundreds during the year that is past. And we are going to sell thousands this year. We have been returned at the head of the poll, as the **cheapest** and **best Lathe makers in the country.** Do you know we send these little Lathes **all over the world.** Protection or no Protection, we command a sale. Why? Because **we give value,** and no country can give more. Send four stamps for our Green Book, and **note the price and quality** of our goods.

THE POPULAR CANDIDATE FOR TOOLS.

S. HOLMES & CO., Engineers,
BRADFORD.

The "PORTALATHE."

The Little Lathe with the Big Cut.

THE Makers beg to introduce to all Amateur Mechanics and Model Makers the "**PORTALATHE**," which is, as its name implies, a Portable Lathe designed to meet the general demand for an extremely rigid and sturdy machine at a popular price.

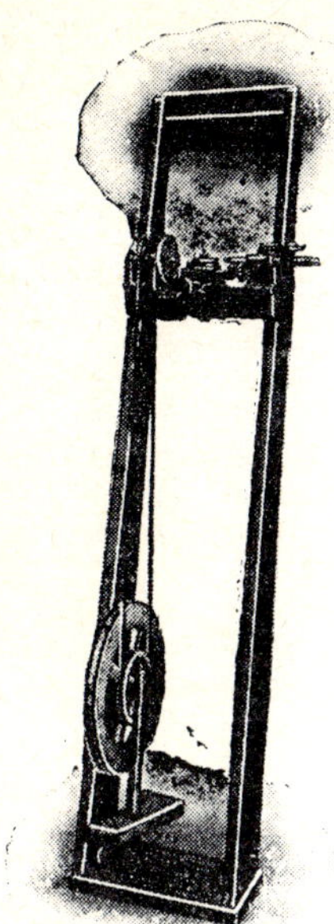
THE "PORTALATHE."

Points Worth Considering:

Absolute rigidity and portability.
No fastening to the floor or wall required.
Lathe can be attached at any working height.
Special Compound Slide Rest will take Heavy Cut.
First-class workmanship and finish throughout.
Perfect interchangeability of parts.
Convenient for storage when not in use.

The Amateur's Lathe on which a really good job can be done.

SPECIFICATION:

Height of centres, 2 in.
Height of centres from bottom of gap, 2⅝ in.
Length between centres, 6 in.
Extreme width, 12 in.
Diameter of face plate, 3¾ in.
Tailstock spindle fitted with lever for drilling.

Special compound slide rest (*prov. patented*), with 2 in. travel on bottom slide and 1¾ in. on top.
Heavy flywheel and footmotor.
Lean-to lathe stand (*fully patented*), fitted with back tray and tool rack.
Finished in grey and blue.

Price £3 : 15 : 0 Complete.

WRITE FOR PARTICULARS OF CHUCKS, ETC.

Manufactured by **The Heeley Motor and Manufacturing Co.,**

Telegrams: "Portal, Sheffield." BROADFIELD ROAD, HEELEY, SHEFFIELD. *Telephone: Sharrow* 53.

The Hughes "DAMACO FIVE"

2½″ Centre, Self-acting, Sliding, Boring and Milling LATHE.

COMPLETE, as illustrated,

£7 : 10 : 0

Free on rail London.
PACKING INCLUDED.
Including Compound Slide Rest, Change-Speed Gear-Box, Lever Scroll Chuck with 2 Sets of 3 Jaws, Driver Chuck, 2 Steel Centres, Drilling Apparatus, Set of 3 Tools; ready for use; also Treadle Attachment.

SCREW-CUTTING MODEL

with 14 Change Wheels, to cut 6 to 120 t.p.i., including metric.

Right and Left Hand.

£10 : 0 : 0

COMPLETE as above.

FITTED FOR HAND, TREADLE, AND POWER DRIVE.

Write us for a descriptive pamphlet of the "Damaco Five."

A real machine tool of "Precision" class.

Your Local Dealer can supply you.

W. E. HUGHES & CO., LTD., 50 & 51, Lime Street, LONDON, E.C.3.

HUGHES, FAWCETT & CO.,

Telegraphic Address
"Hughes, Fawcett
Hebden Bridge.
Telephone:
No. 17 H.B.

Engineers, Tool-Makers, Ironfounders,

Head Office and Works: Hebden Vale Iron Works, **HEBDEN BRIDGE** (Eng.).

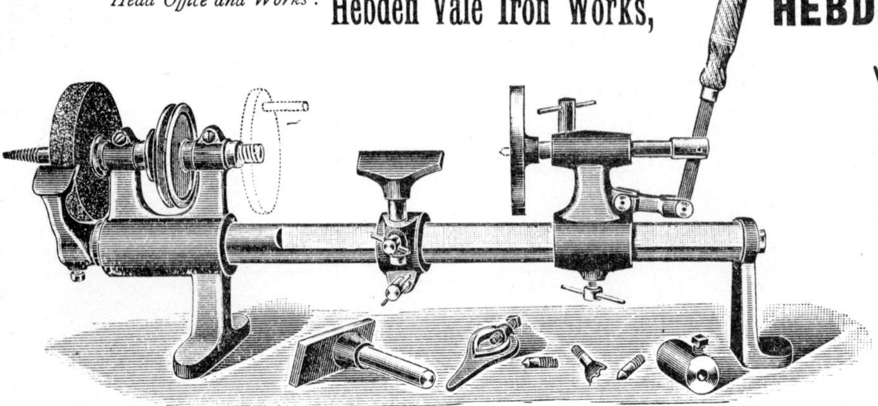

The "Versatile" Combination Lathe (Lathe, Drill Emery Grinder, Polishing Head).

Liverpool Depôt:
34, PARK LANE.

Manchester Depôt:
272a, DEANSGATE.

Catalogue Free all over the World.

Terms: CASH or APPROVED REFERENCES or, "The Times" System of Easy Payments.

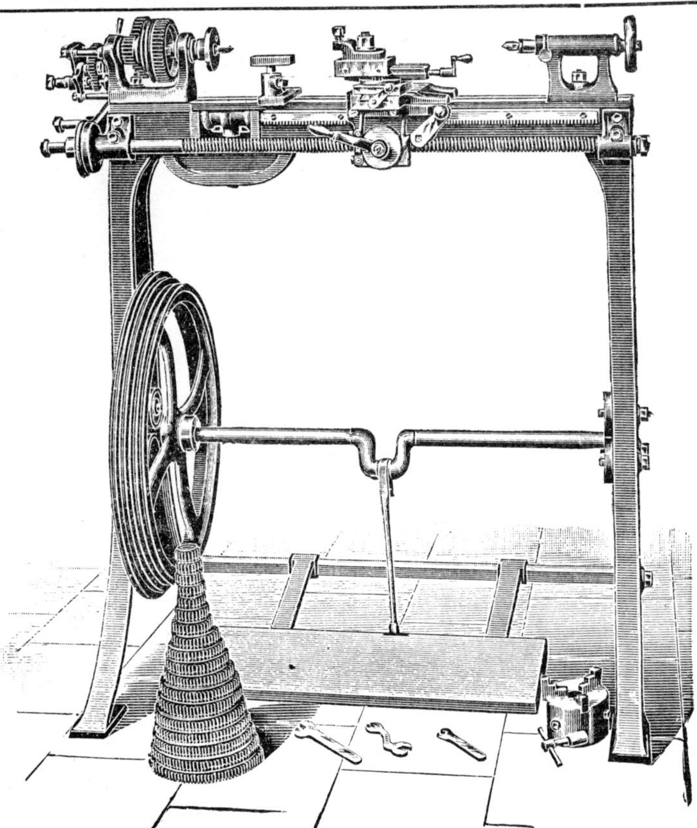

The "Huforko" Lathes. High-Class Machines only. All Sizes from 2½ centres.

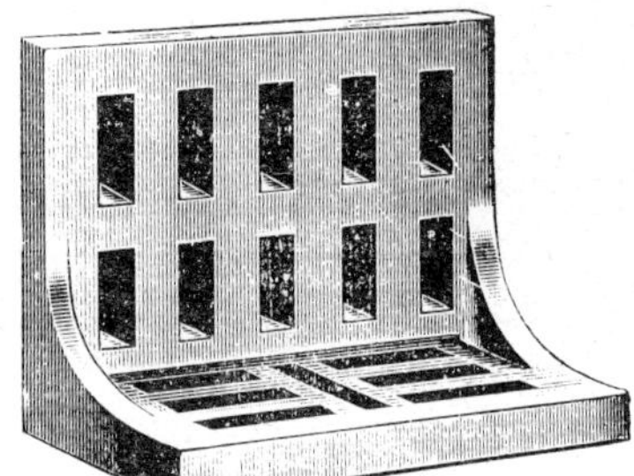

Angle Plates (144 Patterns and Styles)

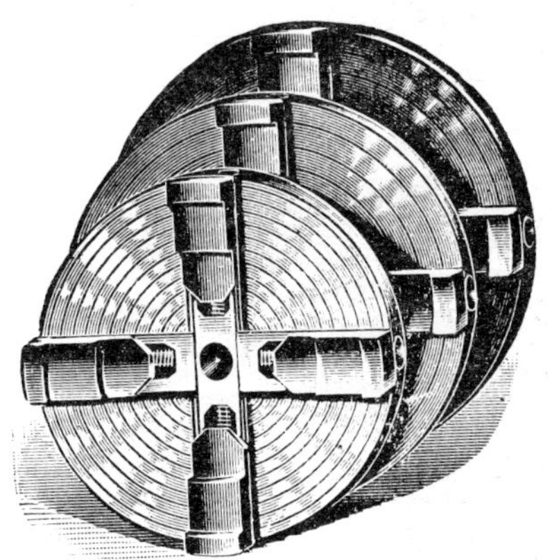

4-Jaw Independent Chucks, from 4 in. to 4 ft. diam

Castings and Forgings Supplied.

WE MEAN BUSINESS. **We are ALWAYS bringing out new things. Keep in touch with us.**

H. T. H.

$2\frac{3}{4}$-inch Precision SCREW-CUTTING LATHE,

With Compound Slide Rest. Complete on Painted Stand, with One Drawer and Foot Treadle. Ready for use.

£16 0 0

SAME LATHE ON OAK STAND, as illustrated, **£20 15 0.**

5" Screw-Cutting Lathe also made.

$2\frac{1}{2}$-inch PLAIN LATHE

for Metal, Wood or Ivory Turning. With Compound Slide Rest. Complete on Stand, with Foot Treadle.

£8 10 0

HUMPAGE, THOMPSON & HARDY, *HIGH-CLASS MACHINE TOOL MAKERS,* **BRISTOL.**

THE 'GEM' LATHE, 24/-.

3" Centre, 9" between centres, screwed nose with driving chuck, plained bed mounted on two small standards, with hand wheel. Back centre, hand-rest complete.

3-speed Foot Motor to suit the above, 19/6.

J. JONES, 44a. Haymerle Road, PECKHAM, S.E.15.

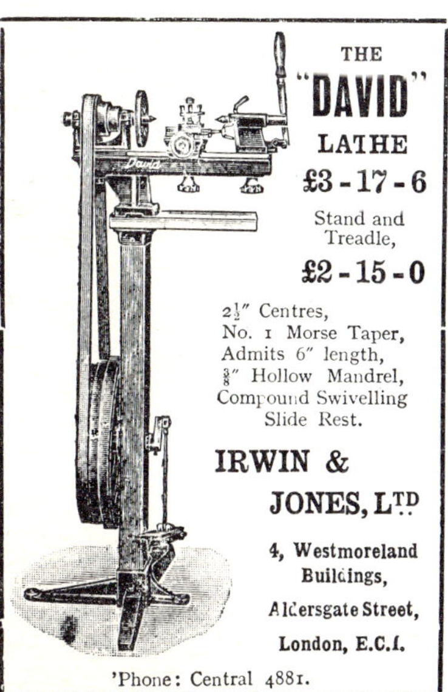

THE

"DAVID" LATHE

£3-17-6

Stand and Treadle,

£2-15-0

$2\frac{1}{2}$" Centres,
No. 1 Morse Taper,
Admits 6" length,
$\frac{3}{8}$" Hollow Mandrel,
Compound Swivelling Slide Rest.

IRWIN & JONES, LTD

4, Westmoreland Buildings, Aldersgate Street, London, E.C.1.

'Phone: Central 4881.

"EMPIRE" Machine Vices.

Indispensable for Small Work.
Holds any shape perfectly rigid.

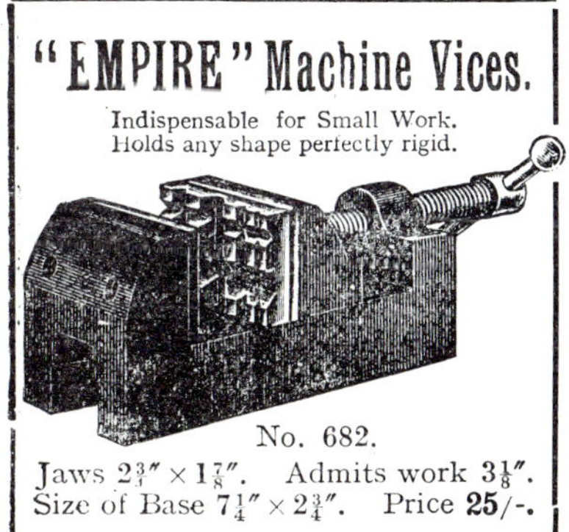

No. 682.

Jaws $2\frac{3}{4}" \times 1\frac{7}{8}"$. Admits work $3\frac{1}{8}"$.
Size of Base $7\frac{1}{4}" \times 2\frac{3}{4}"$. Price **25/-.**

No. 681.

Jaws $2\frac{3}{8}" \times 1\frac{3}{8}"$. Admits work $2\frac{7}{8}"$.
Size of Base $6" \times 2\frac{3}{8}"$. Price **12/-.**

Jones & Shipman, Ltd., Leicester,

and 163, King Street East, TORONTO.

Be up-to-date by using

J. & S. TURNING-TOOL HOLDERS

STRAIGHT SHANK.

RIGHT HAND OR LEFD HAND.

with High-speed Steel Tool Blades ground ready for use.

Made to suit **Drummond, Edgar, Britannia, Bantam Lathes, etc.**

Also stocked in larger sizes for lathes up to 30-inch centres.

A. A. JONES & SHIPMAN, LTD., LEICESTER.

SHEFFIELD STEEL.

Get it direct from the furnace and the forge. Genuine Sheffield-made from start to finish. Guaranteed quality at best current prices. The very pick of the world's steel produce at manufacturer's cost. Every variety for every purpose. All sizes—rounds, squares, and sections. Small lengths supplied. Send for list.

Best Cast Tool Steel	**Magnet Steel**
High Speed Steel	**Die Steel**
Self-hardening Steel	**Drill Steel**
Bright Mild Steel	**Bessemer Steel**

SHEFFIELD TOOLS.

Inman's First Quality Milling Cutters, Reamers, Taps, Dies, Die Nuts, Drills, Hacksaws, Files, Marks and Letter Punches. Guaranteed Precision Tools, Gauges, etc.

INMAN'S Famous Lathe Tools. Guaranteed.

SHEFFIELD MADE.

Price per Set, Best Cast Steel: 1/4" 7/-, 3/8" 9/-, 1/2" 11/-. Sample 2/-. **High-Speed Self-hardening Steel,** price double. Finest tools made. Any size or shape made to order. Every Tool Branded.

INMAN'S Master Tool Holder. New Design, Single Nut Adjustment.

SHEFFIELD CASTINGS

and Forgings, etc., to your own sketches or patterns, in iron, steel, brass and aluminium. **PATTERNS SUPPLIED.** Lowest prices. Satisfaction guaranteed. All sizes Backplates, Faceplates, Drilling Pads, Pulley Wheels, etc., stocked.

ACETYLENE WELDING AND MACHINING of all kinds. Crankshafts turned. Gears cut, etc.

INMAN & Co.,

Upperthorpe, SHEFFIELD.

KEATS' PATENT

VEE ANGLE PLATE

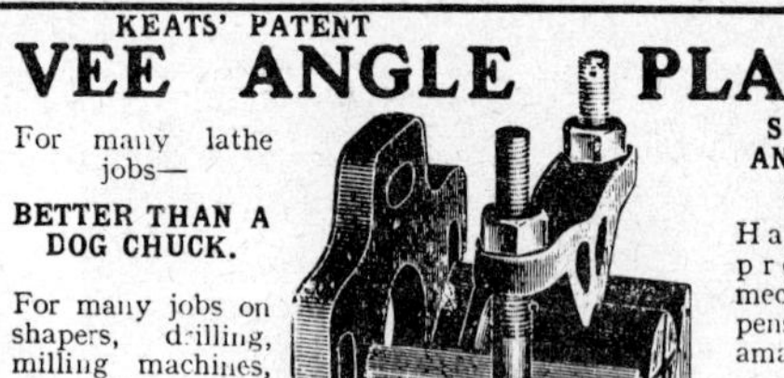

For many lathe jobs—

BETTER THAN A DOG CHUCK.

For many jobs on shapers, drilling, milling machines, etc.—

BETTER THAN A MACHINE VICE.

Costs much less than either.

SAVES TIME AND ENSURES ACCURACY.

Handy to the professional mechanic — indispensable to the amateur.

Supplied to many leading firms

Four standard sizes to suit lathes 3" centres upwards.

THE TOOL FOR THE AWKWARD JOB.

See it demonstrated at M.E. Exhibition, Stand No. 2, or write for particulars.

KEATS, 4, SHELFORD ROAD, MILTON, PORTSMOUTH.

Leabrooks Engineering Co

Near ALFRETON.

SEND FOR LISTS.

S.S.S. Precision Lathes and Machine Tools.

30/- VALUE for 11/6

SPECIAL ONE-WEEK SALE OF BENCH DRILLS

Manufacturer, forced by crisis to obtain cash at once, is sacrificing these high-grade **30/- Bench Drills** at **11/6** (carr. 1/- extra) for 7 days only! Seize this opportunity! **Cut gears, Steel Spindle, Ball Bearings, automatic adjustable feed, fitted with 3-jawed chuck. Capacity 0 to 3/8 in.** Will last lifetime. **SET OF DRILLS 1/- extra if required.** Special offer of **BREAST DRILLS,** 0 to 3/8 in., enclosed gears. Usual price 15/6—to clear at **6/6** carr. paid. **BENCH GRINDERS,** 4 in. wheel, **3/6** post free. 6 in. wheel, **7/-** post free.

Money back if not delighted. Bargain list free.

JONES' STORES (Dept. M. E.,) 51, London Rd., Croydon

PRICE
£9 15s.

Self-Acting Bench Shaping Machine,

— FOR —

POWER OR HAND.

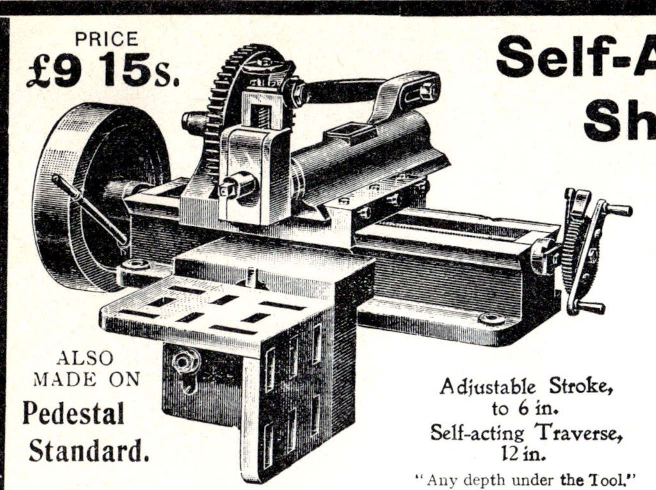

ALSO MADE ON
Pedestal Standard.

Adjustable Stroke, to 6 in.
Self-acting Traverse, 12 in.
"Any depth under the Tool."

A Diesinker, working in **steel**, writes: "Shaper is giving every satisfaction."

A Brassfinisher, working in **brass**, writes: "We are exceedingly well pleased with the machine."

An Engineer, working in **cast iron**, writes: "I am still using the Bench Shaper bought from you over two years ago. It still seems as good as ever."

Spindles of Large Diameter.
Gears Cut from the Solid.
Large Bearing Surfaces.

Suitable for all Kinds of Light Work.

Vices, Tools, Keyseating Attachments, Angle Brackets, &c., extra.

Sole Agents for Scotland :—
The Fairbanks Company,
54-56, BOTHWELL STREET,
GLASGOW.

SOLE — MAKERS, LEYLAND BARLOW & CO.,
TRAFFORD PARK, MANCHESTER.

SELF-ACTING SHAPING MACHINE, £12

LEYLAND BARLOW & CO.,
SOLE MAKERS,
TRAFFORD ROAD, MANCHESTER.

HAND SHAPING MACHINE

Stroke to 6 in.
Self-acting Traverse 8 in.
Will slot shaft to 2 in. diameter.
Takes any depth under the Tool.

Price - - £6 15s.

VICES, TOOLS, EXTRA.

LEYLAND BARLOW & CO.,
SOLE MAKERS,
Trafford Road, Manchester.

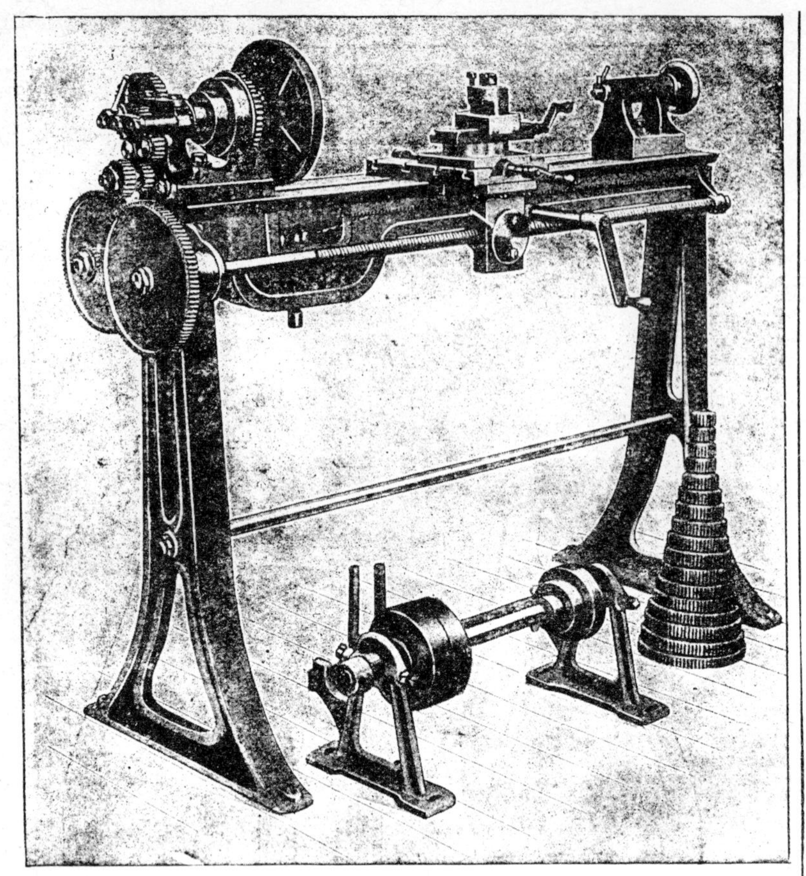

Made by LEYLAND BARLOW & Co., Trafford Rd., Manchester

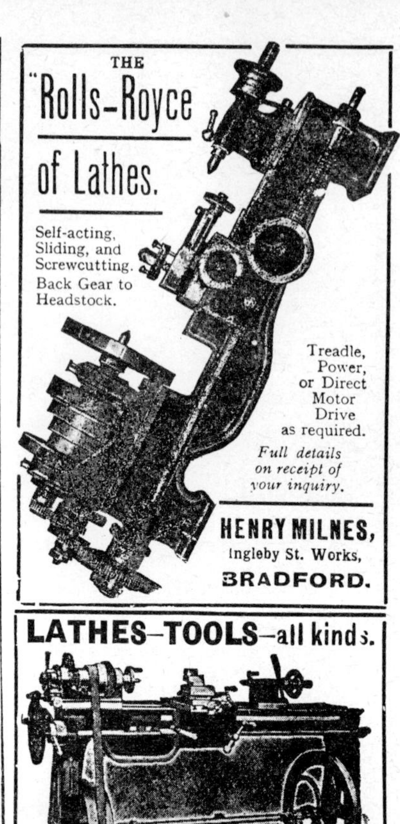

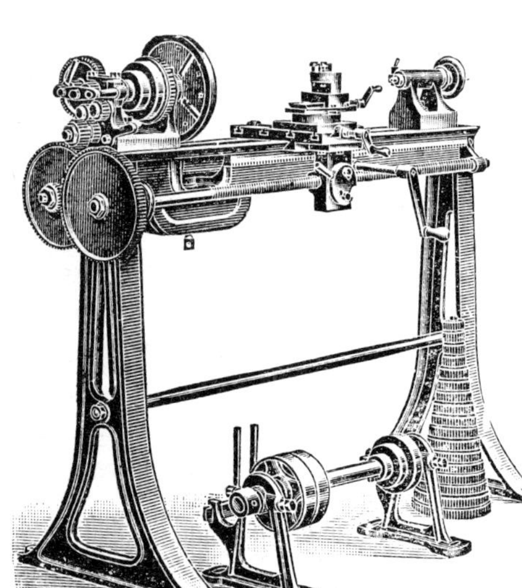

4½ ins. centre
4 ft. gap bed

Screw-Cutting Lathe.

For Power and Foot.

Price List, 1d. Stamp.

LEYLAND BARLOW & Co. Makers, TRAFFORD RD., MANCHESTER.

THE LITTLE BEATALL

3 ins. centres, 24 ins. bed.
Three hand-turning tools, and centres all hardened, ready for use.

IS THE FAVOURITE. 17/6

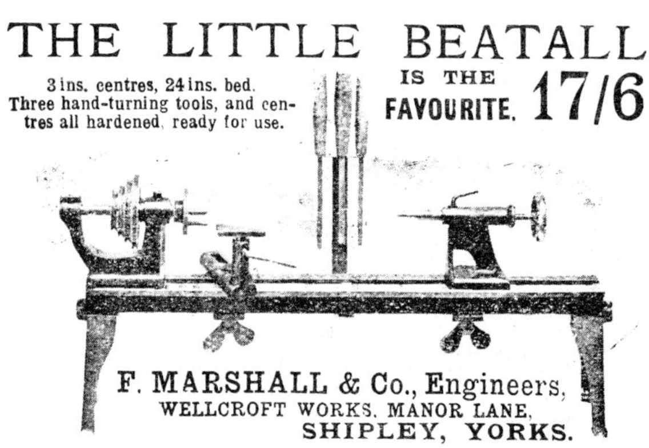

NOTE.—WE CAN SUPPLY a 4-JAWED CHUCK and COMPOUND SLIDE REST WITH THIS LATHE IF REQUIRED.

Rd. MELHUISH, Ltd.

50, 51, 84, Fetter Lane, Holborn, E.C.

Branch Est.

143, HOLBORN BARS.

We keep always for immediate dispatch a large supply of Tubing, Brass, Copper, and Steel, as used by Model Engineers. B.A. & Whitworth Screws

THE MODEL ENGINEER'S
HAND PLANER.

This is of excellent build and finish, and well within the means of the ordinary mechanic. The tool box is of different design to those usually supplied, and enables the tool to be placed in any position.

Length of Bed, 14 ins. Width, 6 ins.
Will take in 5 ins.
Approximate Weight, 100 lbs.

Price, including two Spanners,

£6 : 0 : 0

(as illustrated).

Machine Vice, suitable for same, **12/6** extra.

Melhuish's

DOUBLE-SPEED PLANER.

New design.

PRICE .. **£13 10s.**

TOOLS # XMAS GIFTS **TOOLS**

THE VERY LATEST IN PRACTICAL LATHES.

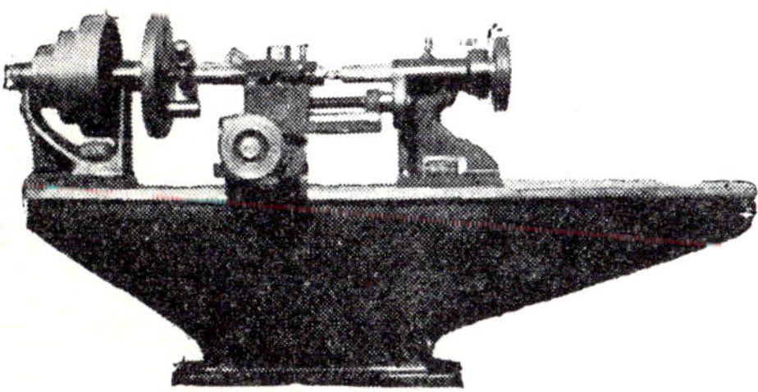

Price only **£5 : 10**

Carriage paid in London radius. Wonderful value !
4½″ centres, 30″ bed, admits 18″ between centres.
Complete with compound slide-rest.

A large assortment of FRETWORK MACHINES, CARVING TOOLS, MATERIALS for BENT-IRON WORK, SHOEMAKER'S TOOL CHESTS, Etc., in Stock.

Money refunded if goods do not meet with satisfaction.

YOUTH'S TOOL CHEST, 1432W.

TOOL CHESTS FOR BOYS.

No. 1432W, containing	7 Tools	...	6/6	Postage	9d.
No. 1433W ,,	10 ,,	...	9/-	,,	1/-
No. 1434W ,,	13 ,,	...	12/6	,,	1/-
No. 1435W ,,	10 ,,	...	17/6	,,	1/-
FOR YOUTHS.					
No. 1436W ,,	19 ,,	...	40/-	carriage	paid.
No. 1438W ,,	26 ,,	...	60/-	,,	,,
No. 1439W ,,	34 ,,	...	80/-	,,	,,
HOLBORN TOOL SET,					
Containing 13 first quality Tools.		...	27/6	carriage	paid.

OUR ILLUSTRATED METALWORKERS' CATALOGUE,
containing 560 pages, sent post free **1/6.**
ILLUSTRATED WOODWORKERS' CATALOGUE,
containing 338 pages, sent post free **1/3.**

RICHARD MELHUISH, LTD.,

50-51, FETTER LANE, HOLBORN CIRCUS, E.C.4.

Branch Depot : (CUTLERY) 143, HOLBORN. **Telegrams : " WORKFOLK, LONDON."** **Telephone : HOLBORN 1611 (3 lines).**

HENRY MILNES,

MANUFACTURER OF

High-Class Lathes

FOR

Sliding, Surfacing, Screw Cutting, Plain and Ornamental Turning.

Treadle Milling Machines Hand Planing Machines, Slide Rests, Chucks, &c. of all descriptions

INGLEBY WORKS, BROWN ROYD, BRADFORD.

ESTABLISHED 1858.

Lathes and Tools of my make may be inspected by any intending purchaser in London or Glasgow, at address given below.

Agents for London—Messrs. E. PADFIELD & CO., West India House, 96 & 98, Leadenhall Street, LONDON, E.C.

Agents for Scotland—Messrs. THOS. HILL & Co., 66, Robertson Street, GLASGOW.

HENRY MILNES LATHE AND TOOL MAKER

High-class Lathes

For Sliding, Surfacing, Screw Cutting, Plain and Ornamental Turning.

Treadle Milling Machines, Hand Planing Machines, Slide-rests, Chucks, &c., of all descriptions.

Ingleby Street Works, Bradford

Est. 1858.

Lathes and Tools of my make may be Seen at THE MODEL ENGINEER Laboratory.

HENRY MILNES,

Lathe and Tool Maker.

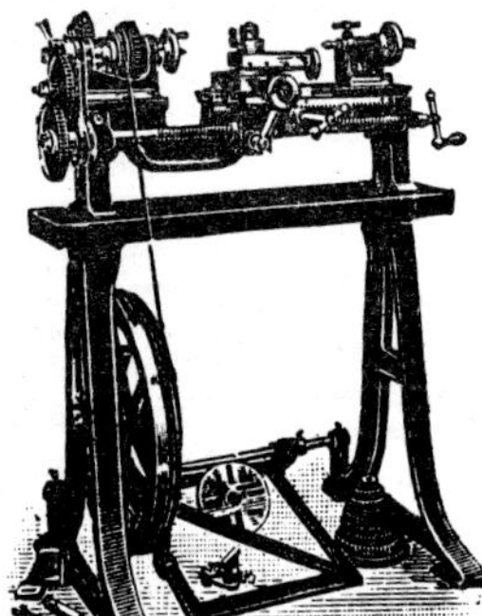

A High-Class Sliding, Surfacing, and Screw-Cutting Lathe for

£16 : 5 : 0

fitted with Boring Table, Compound Slide Rest, Hollow Mandrel, Treadle Motion or Countershaft as required, Machine Cut Gears throughout.

Accuracy Guaranteed.

Ingleby Street Works, **BRADFORD.**

Lathes and Tools of my make can be seen at "THE MODEL ENGINEER" Laboratory.

HENRY MILNES,

Manufacturer of **High-class Lathes**

— FOR —

Sliding, Surfacing, Screw Cutting, Plain and Ornamental Turning.

Treadle Milling Machines, Hand Planing Machines, Slide Rests, Chucks, &c. of all descriptions.

Ingleby Works,

BROWN ROYD, BRADFORD

ESTABLISHED 1858.

Lathes and Tools of my make may be inspected by any intending purchaser at London or Glasgow at address given below.

Agents for London—Messrs. E. PADFIELD & CO., West India House, 96 & 98, Leadenhall Street, LONDON, E.C.

Agents for Scotland, Messrs. THOS. HILL & Co., 66, Robertson Street, GLASGOW.

HENRY MILNES,

Lathe and Tool Maker.

A High-Class Sliding, Surfacing, and Screw-Cutting Lathe, fitted with Boring Table, Compound Slide Rest, Hollow Mandrel Treadle Motion or Countershaft as required, Machine Cut Gears throughout.

Accuracy Guaranteed.

Ingleby Street Works, **BRADFORD.**

Lathes and Tools of my make can be seen at "THE MODEL ENGINEER" Workshop.

Lathes can only be supplied when requiredfor War Service.

HENRY MILNES,

Manufacturer of HIGH-CLASS LATHES

For Sliding, Surfacing, Screw Cutting, Plain and Ornamental Turning.

Treadle Milling Machines, Hand Planing Machines, Slide Rests, Chucks, etc., of all descriptions.

INLGEBY WORKS, Brown Royd, BRADFORD.

**

Established 1858.

**

Lathes and Tools of my make may be inspected by any intending purchaser at London or Glasgow at address given below.

Agents for London—Messrs. E PADFIELD & CO., West India House, 96 & 98, Leadenhall Street, LONDON, E.C.

Agents for Scotland, Messrs. THOS. HILL & Co., 66, Robertson Street, GLASGOW.

UNEXCELLED FOR QUALITY OF WORKMANSHIP AND MATERIAL.

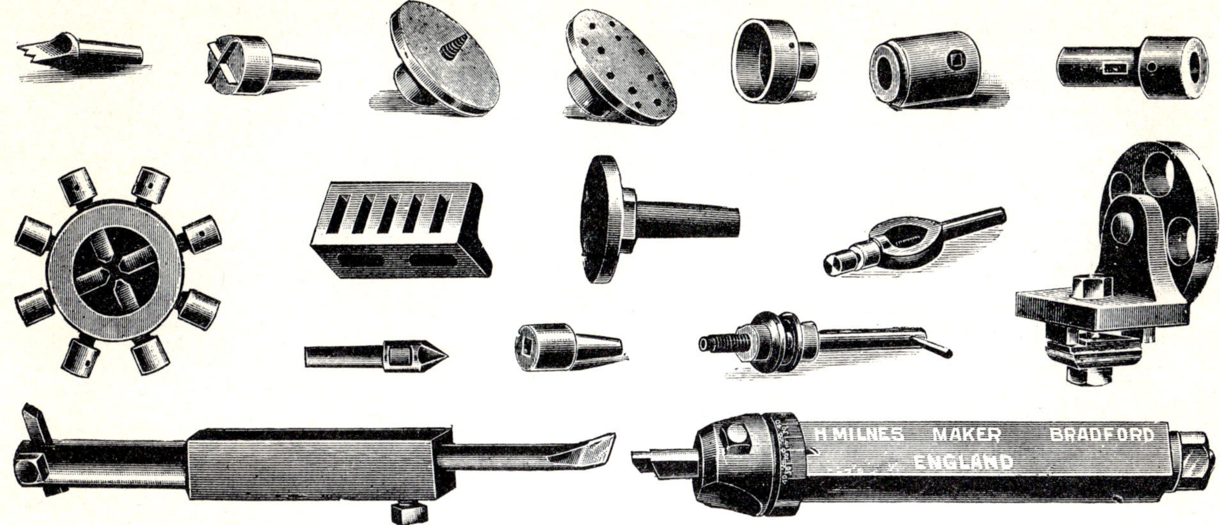

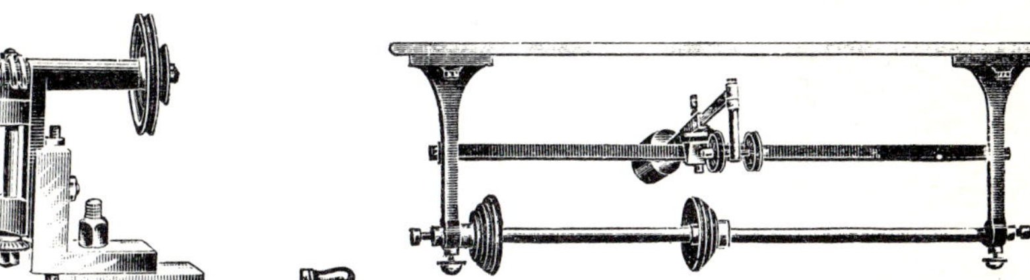

Special Cutter Bar for Boring and Internal Screw-Cutting.

Special Cutter Bar for Screw-Cutting.

Metal Spinning Tools.

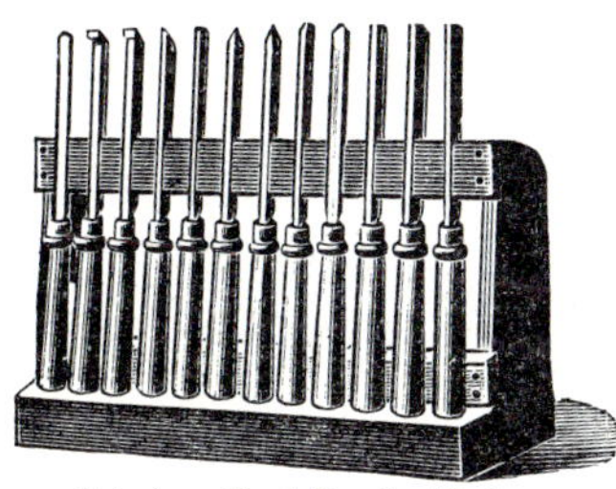

Slide-rest Attachment for Wheel-Cutting, Tap Grooving, Milling.

Overhead Apparatus.

Set of 12 Slide-Rest Tools.

Set of 12 Hand-Turning Tools.

Set of 13 Flat Drills, with Holder. Will bore holes from ⅛ in. up to 1 in. dia.

Lathes and Tools of my make may be inspected by any intending purchaser in London or Glasgow, at addresses given below.

AGENTS FOR LONDON—

Messrs. E. PADFIELD & Co., West India House, 96 & 98, Leadenhall Street, LONDON, E.C.

AGENTS FOR SCOTLAND—

Messrs. THOS. HILL & CO., 66, Robertson Street, GLASGOW.

Drilling Spindle No. 2.

EVERYBODY WHO HAS A LATHE should send for my complete Illustrated Catalogue of 104 pages. Contains particulars of all kinds of lathes, drilling machines, ornamental turning appliances, chucks, tools, cutter-bars, fret-saws, cutter-frames, wheel cutting appliances, drills, grindstones, gauges, lathe castings, and parts, hand-planing machines, and milling machines and cutters. Also drawings of a ½ h.-p. horizontal steam engine, for which I supply castings. The list is beautifully printed on art paper, and will be sent on receipt of stamps value 6d. to cover postage to any reader mentioning THE MODEL ENGINEER.

HENRY MILNES, HIGH-CLASS LATHE and TOOL MAKER, Ingleby Works, Bradford.

UNEXCELLED FOR QUALITY OF WORKMANSHIP & MATERIAL.

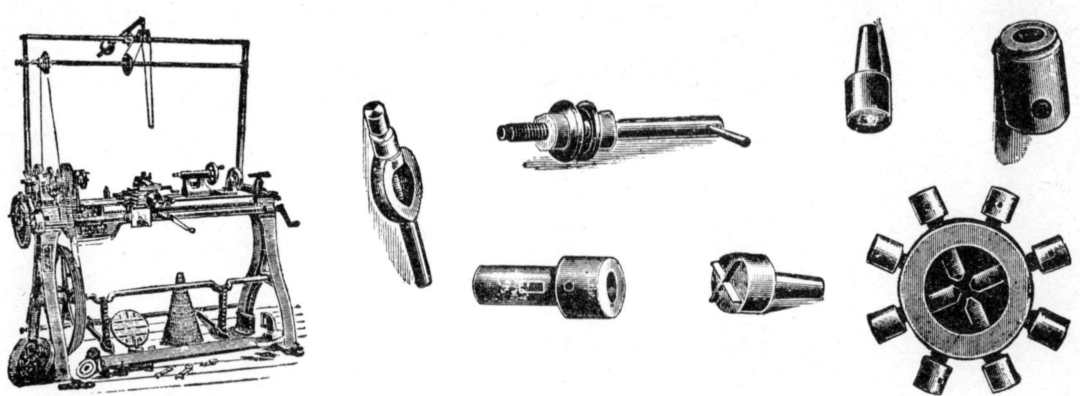

AGENTS FOR LONDON—Messrs. **E. PADFIELD & Co., West India House, 96 & 98, Leadenhall Street, LONDON, E.C.**
AGENTS FOR SCOTLAND—Messrs. **THOS. HILL & Co., 66, Robertson Street, GLASGOW.**

EVERYBODY WHO HAS A LATHE should send for my complete Illustrated Catalogue of 104 pages. Contains particulars of all kinds of lathes, drilling machines, ornamental turning appliances, chucks, tools, cutter-bars, fret-saws, cutter-frames, wheel cutting appliances, drills, grindstones, gauges, lathe castings, and parts, hand-planing machines, and milling machines and cutters. Also drawings of a ¼ h.-p. horizontal steam engine, for which I supply castings. The list is beautifully printed on art paper, and will be sent on receipt of stamps value 6d. to cover postage to any reader mentioning THE MODEL ENGINEER.

HENRY MILNES, HIGH-CLASS LATHE and TOOL MAKER. **Ingleby Works, Bradford.**

See my Exhibit, STAND No. 163, near the Arcade Entrance, Stanley Show, November 19th to 27th.

COMBINATION LATHE, Type "Y."

HENRY MILNES, Ingleby St. Works, BRADFORD.

"LATHES of QUALITY"

4⅛″ Centre, 4′ 0″ Gap Bed.
Sliding, Surfacing and Screwcutting.
Stop Motion to Longitudinal Feeds.
Dial Indicator for Screwcutting.

HENRY MILNES, Ingleby St. Wks., **Bradford**

Real Lathes these "Milnes."

Full details await your P.C.

Henry Milnes, Ingjeby St. Works, BRBDFORD.

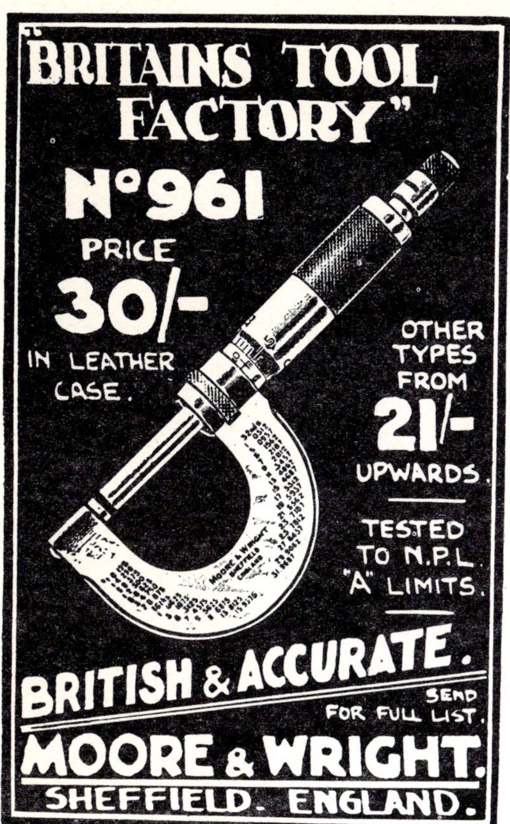

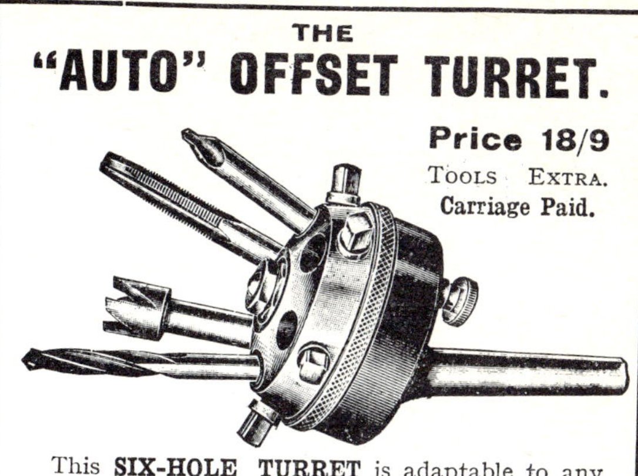

This **SIX-HOLE TURRET** is adaptable to any lathe having a No. 1 Morse Taper in the tailstock. It is so designed that the tools which are not in use swing well clear of the work, and the tool which is in use comes in direct line with the taper shank, thus eliminating setting-up difficulties.

The **Turret Head** is of **Steel,** and revolves in an accurately-machined recess, being held there by a washer and two lock-nuts. The **Tools** automatically centre and lock themselves in position.

The Whole Fixture is strong, accurate, and well-finished, and fit for any high-class lathe.

Call at Stand No. 1 at the MODEL ENGINEER EXHIBITION and inspect it, also the N.W. 2⅜″ Lathe, Treadle and Power Drilling Machines.

NICHOLLS BROTHERS, BRAINTREE, ESSEX.

LATHES FOR AMATEURS.
LATHES FOR MODEL MAKERS.

This Block represents our 4 in. with 2 ft. 6 in. Bed.
Price Complete, £15.

F. MITCHELL & CO., Victoria Works,
Longley Road, Tooting Junction, S.W.

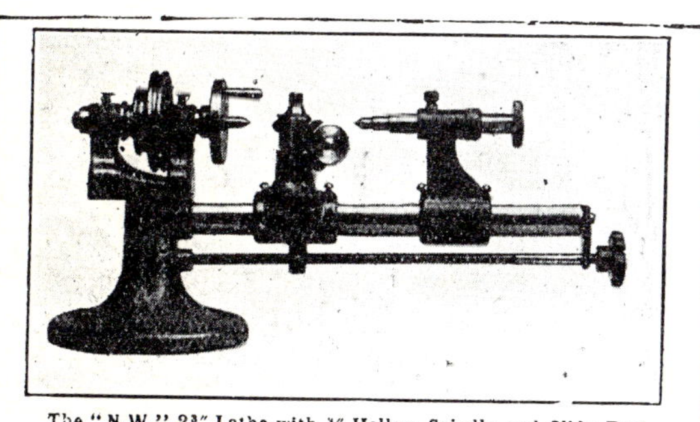

The "N.W." 2⅜″ Lathe with ⅜″ Hollow Spindle and Slide Rest.
Price £2 15 0. Stamp for particulars.
NICHOLLS BROTHERS, BRAINTREE, ESSEX

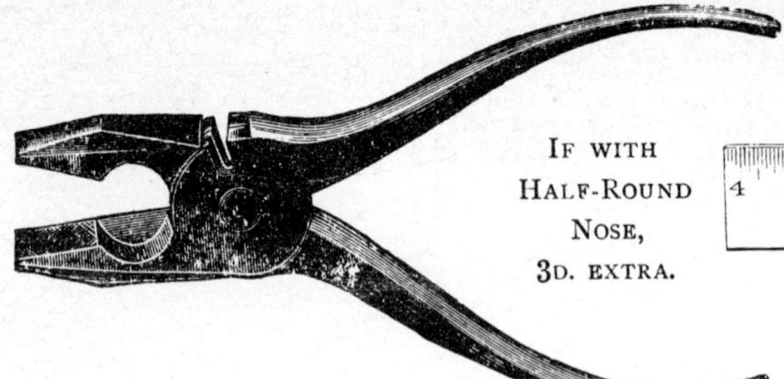

IF WITH HALF-ROUND NOSE, 3D. EXTRA.

REALLY GOOD FLAT NOSE CUTTING PLIERS.

5½ in 1/6, 6 in. 1/9, 7 in. 2/-, 8 in. 2/9, per pair, post free.

CHESTERMAN'S Warranted STEEL SQUARES. . . .

CHESTERMAN SHEFFIELD ENGLAND No 1371

	3 in.	4 in.	6 in.
Plain . . .	4/.,	4/6,	7/6
Rule Marked,	5/-,	6/-,	10/-

POST FREE.

ACME . . . DRILL CHUCK,

Holds 0 to ½ in.

13/6, post free.

Don't forget to send for our No. 10 Catalogue, 4d. Post Free.

No 10.

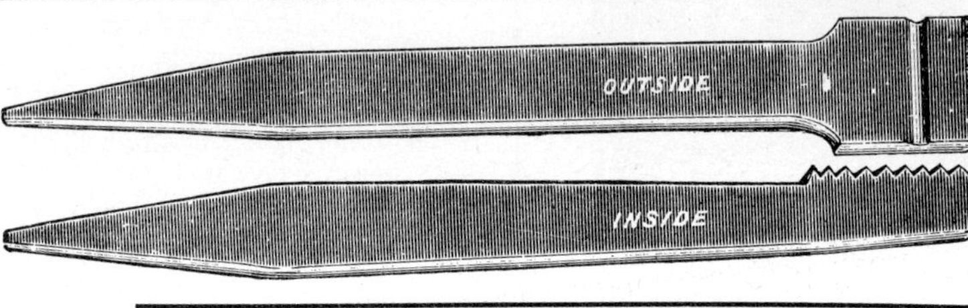

ORDINARY . . . SCREW TOOLS.

14 Threads per inch, and all finer Threads.

1/6 per pair, post free.

CHARLES NURSE & CO., Tool Merchants,

182, 184, and 173, Walworth Road, LONDON, S.E.

To Drummond 4-inch Lathe Users.

We supply a table of correct change-wheels for screw-cutting specially worked out to suit this lathe. Neatly printed on thin card for workshop use. Price 6d., post free.

PERCIVAL MARSHALL & Co.,

66, FARRINGDON STREET, LONDON, E.C.4.

Build Your Own Independent Chuck.

COMPLETE SET OF MATERIALS includes: body casting, forgings for jaws and key, steel for screws, and blue print.

4" diam. **4/-** per set, post 9d.
5" „ **5/-** „ „ „ 1/-.
6" „ **6/-** „ „ „ 1/3.
8" „ **8/-** „ „ carr. fwd.
10" „ **10/-** „ „ „ „

Planing out grooves for jaws 4", 5", 6", **2/-** extra.
Ditto. 8" and 10", **3/-** „
Screwing out Whit. threads, 4", 5", 6", **2/6** „
Ditto ... ... 8" and 10", **3/6** „
Special threads, **1/6** extra.

My 52-page catalogue illustrates a host of useful machines, attachments and castings for the amateur engineer, 6d. post free by return mail.

TOM SENIOR,

' Atlas ' Works, Hightown, LIVERSEDGE, Yorks

POLISHING HEAD, 6/-.

TOOL BARGAINS.

WALWORTH SYSTEM PIPE STOCKS AND DIES. ⅛", ¼", ⅜", ½", 15/-; ½", ¾", 1", 15/-.

WHITWORTH STOCKS AND DIES. ⅛", 3/16", ¼", **7/6**; with Taper and Plug Taps to each size.

DREADNOUGHT FILES. 6" **9d.**, 8" **1/-**, 10" **1/3**, 12" **1/6**, 14" **1/10**.

Special Line of TOOL CHESTS.

Inside capacity 21" × 11½" × 7½", strongly made, with banded corners, painted inside and out. Cannot be repeated at the price. Only **3/-** each.

JOHN OSBORN, 100, FLEET ST., E.C.4 (Adjoining Ludgate Circus).

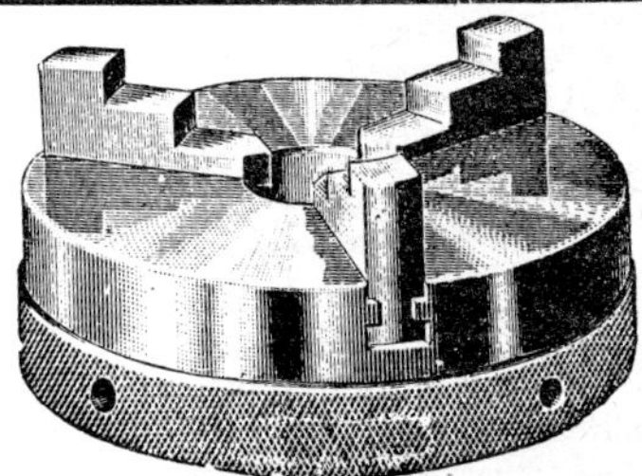

2" **Lever Scroll Chuck, 37/6.** 3", **40/-.** With two Sets of Jaws.

SPECIAL! 2¾" Cushman Patt. CHUCK, Two Sets of Jaws, **22/6.**

PATRICK LATHES.

2″ Model.

The Wonder Model 2 in., 3 in., 4½ in. Patrick Lathes still represent the best value that YOUR money can buy. Hollow Spindles, Screwed Mandrel Nose, Taper Centres, Compound Slide Rest.

Ask for Sheet 8c post free.

CASH or
DEFERRED PAYMENTS.

Centres.	Length of Bed.	Admits in Centres.	Admits in Gap.	Single Gear.	Treadle Lathe.	Backgear Extra.
				£ s. d.	£ s. d.	£ s. d.
2″	18″	11″	6″×1¼″	2 15 0	5 5 0	1 10 0
3″	27″	18″	9″×1½″	4 0 0	6 15 0	2 0 0
3″	36″	27½″	9″×1½″	4 15 0	7 15 0	2 0 0
4½″	30″	20½″	11½″×2″	10 2 6	18 2 6	2 17 6

ACCESSORIES.

	Foot Motor Turned.	Three-Point Steady.	Four-Jaw Chuck.	Steel Angle Backplate.	Hand Rest.	C.S. Rest.
	£ s. d.	£ s. d.	£ s. d.	£ s. d.	£ s. d.	£ s. d.
2″	1 10 0	0 8 6	0 15 9	0 3 0	0 5 9	1 3 6
3″	1 10 0	0 11 6	1 1 0	0 3 9	0 8 3	1 11 6

3″ and 4½″ Models.

The "PATRICK," "POPULAR" and "STANDARD"

Models, 4½ in. Screwcutting Lathes still occupy the Premier Position in their respective classes.
Lists 19c and 24c describe these.

Showing Lathe No. 15.
Lathe No. 23 admits of 12 in. more between centres.

Visit Our Stand No. 17 at the Exhibition.

Pack. & Del. C.I.F.	Cape Town	Calcutta.	Ceylon.	N.Z.	Melbourne	Canada.	
EXTRA.	£ s. d.	£ s. d.	£ s. d.	£ s. d.	£ s. d.	£ s. d.	Accessories can be included without any extra cost of carriage.
2″ Bench	1 2 6	2 7 6	2 16 0	2 11 0	1 10 0	1 8 0	
2″ Treadle	1 10 0	2 7 6	2 16 0	2 11 0	2 0 0	1 12 0	
3″ Bench	1 7 0	2 7 6	2 16 0	2 11 0	1 10 0	1 10 0	
3″ Treadle	1 12 0	2 7 6	2 16 0	2 11 0	2 0 0	1 18 0	

Note New Address:

F. PATRICK, Ltd., SWINNOW LANE, BRAMLEY, Leeds.

"PATRICK"

4″ Screw-cutting Lathe.

Bearings are of Gun-metal and Adjustable. Hollow Spindle, Compound Slide-rest and Boring Table. Machine-cut Change Wheels supplied with each Lathe. Bores Twin and Triple Cylinders at one setting. Every Lathe sold on approval, and money returned if not to your satisfaction.

Send for Sheet No. 15 giving all particulars.

BENCH LATHE .. £5 10 0.
TREADLE „ .. £8 10 0.
BACK GEAR £2 0 0 EXTRA.

MAKER:

F. PATRICK, CROSVENOR TERRACE, LEEDS.

PATRICK 4½″ LATHES

with the ¾″ HOLLOW Spindle.
Standard and Popular Models. First-Class Workmanship. Ask for Lists 15C or 14C.
Deferred Payments Arranged.

F. PATRICK, Ltd.,
SWINNOW LANE, BRAMLEY, LEEDS.

Taps 6d. Dies 1/-.

Sizes in Whitworth: 1/16″, 3/32″, 1/8″, 5/32″, 3/16″, 7/32″, 1/4″; in B.S.F.: 3/16″ and ¼″; in Brass Gas: 3/16″, 7/32″, 1/4″; in B.A.: 0, 1, 2, 3, 4, 5, 6, 7, 8, 9, 10; in Model Engineer: ⅛″, 5/32″, 3/16″, 7/32″, ¼″. Prices for Taps in above List, **6d.** each; 5/16″ Taps, **9d.**; ⅜″, **1/-**; 7/16″, **1/3**; ½″, **1/6.** Same price for all standard sizes.

DIES.—13/16″ Circular Adjustable, sizes up to ⅜″, **1/-**; 1″ Circular Dies, **1/6**; Tap Wrenches, **1/-**; 13/16″ Die Stocks, **1/6**; 1″ Die Stock, **2/9.**

Soluble Oil, the best lubricant engineers can use, for parting of or turning metal, **12/-** per gallon, **6/6** half, **3/6** quart.

8½″ **Circular Saws** for wood (second-hand), **2/6,** hole 1″; 6″ as above, **2/-.**

All goods are new except where stated. All orders over 10/- post free.

J. PEACOCK, 114, HIGH STREET, ISLINGTON, N.1.

POLISHING HEAD.

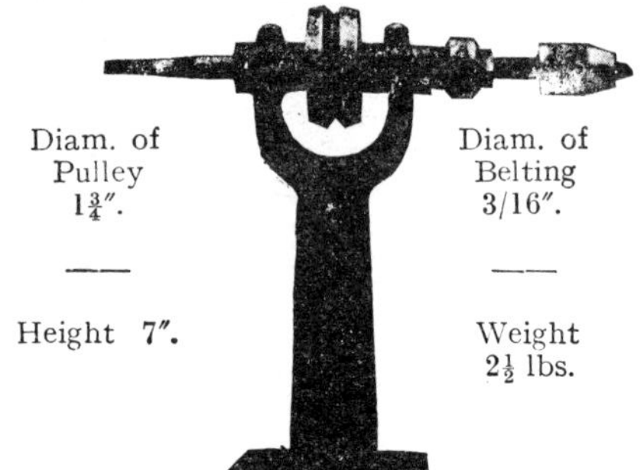

Diam. of Pulley 1¾″.

Diam. of Belting 3/16″.

Height 7″.

Weight 2½ lbs.

A very useful Tool, fitted with 3-jaw Drill Chuck to take Drills up to 3/16″ diam. Spindle is Machined Steel ⅜″ diam., taper-screwed, with collar to take ¾″ thick abrasive wheel.

Price - - 9s.

Similar Head to above, made much heavier, with ¼″ Chuck ... ... **Price 12s.**

UNIVERSAL HEAD.

This little Machine is an extremely useful and versatile Tool, adapted for Polishing, Grinding, Drilling, Sawing, and Milling jobs with ease. Fitted with 3″ Circular Metal Saw, 4″ Abrasive Wheel, Drill Chuck to drill up to ¼″, and Spindle Taper-screwed for Polishing Bobs.

Price only - - 35s. net.

SAMUEL PEACE & SONS, Ltd.,

File and Tool Manufacturers, **21, Red Lion Street, CLERKENWELL, LONDON, E.C.1.**

All Orders and Inquiries will be put through our nearest Agents.

A REAL GOOD LATHE

4½ in. Lathe, complete as above, £10 10s.

Length of bed, 32 ins. Length between centres, 14 ins.

This is a very highly finished, accurate, and practical Lathe.

The workmanship is the best possible, and the design is essentially modern and practical.

The Headstock Spindle is carefully hardened and ground, and runs in hard steel bearings.

The Bed is very heavy and rigid, and is bolted to heavy iron standards.

The Crank-shafts and Flywheel are turned and polished, and run on hardened centres.

The Lathe as above includes treadle and flywheel, standards, polished oak table and tool drawer, eight-screw bell chuck, two-die chuck, poppet cover for drilling, belt, spanners, and keys.

Weight about 300 lbs.

Slide-rest with taper graduation and protected lead-screw ... ... **£3 10s.**

The same Lathe as above, with hollow spindle 10 mm. bore, complete with slide-rest, &c. ... **£16 10s.**

Our Catalogues 1½d. in stamps. Mention this paper.

THE PITTLER CO.,

144, High Holborn, W.C.

You can Depend

Upon us for the RIGHT THING every time.

HERE IS AN INDISPENSABLE TOOL FOR MODEL MAKERS.

ALL STEEL SLIDE GAUGES.

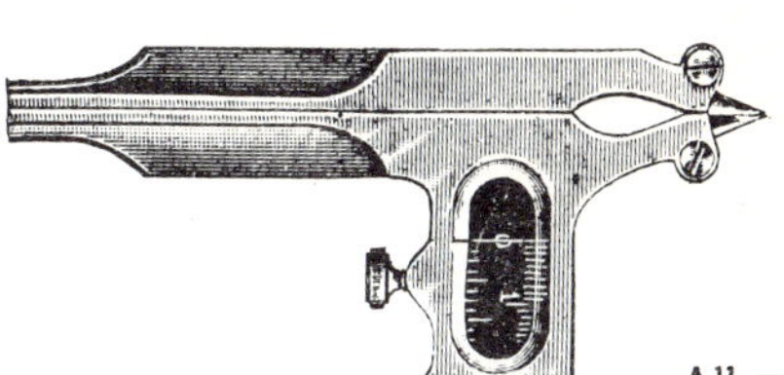

This is No. 71, and, as you see, is a nicely designed tool. It has Cast Steel interchangeable Scribing Points on the back of jaws.

Price - - 5/- each.

Nickelled 5/6 „

No. 73 is like this, but has also a MICROMETER ADJUSTMENT.

Price - - 8/6 each.

Nickelled 9/- „

All are 8 ins long, and have 1¼ ins. jaws.

ENGLISH MEASUREMENTS SUPPLIED ONLY, BUT METRIC CAN BE ADDED FOR 3d. EXTRA.

No. 70 is similar, but without the Scribing Points.

	each.
Price -	**4/-**
Nickelled	**4/6**

We also make a very Precise

POCKET SLIDE GAUGE

5 ins. long, at 4/- each ; Nickelled, 4/3 each

The Pittler Company,

144, High Holborn, London, W.C.

Have you seen the "GRI-POL" COMBN. GRINDER?

The machine to give your model that exhibition finish we all try to get.

It comprises Band Polisher, Grinder, Polishing Head and Twist Drill Grinding Jig which ensures a really accurate job.

No. 1 as above **67/6** No. 2 less Jig **52/6**

SEE IT AT LEICESTER M.E. EXHIBITION—DEC. 14, 15 & 16

POOLS TOOL Co. Ltd.
16 & 18, CARLTON ST., NOTTINGHAM

The "ADEPT" $1\frac{5}{8}$ in. Centre LATHE

Combines beauty of design with real utility. Fitted with fully compound slide-rest. Tailstock has a sliding barrel which is clamped in position by a set screw. Principal dimensions: Height of centres over bed $1\frac{5}{8}$"; height of centres over gap, $2\frac{1}{8}$"; length of work taken between centres, 6"; length over all, $11\frac{1}{4}$".

PRICE: With Compound Slide Rest ... **£1** With Hand Rest instead of Compound Slide **12 6** (Screwed Barrel Tailstock, 5/- extra). Postage 1/-.

Accessories.

3-jaw Dog Chuck to take from 1/16" to ½" diam. ...	4 6
Hand Rest	2 -
Carrier ⅜" diam. ...	1 -
Carrier ⅝" diam. ...	1 3
Faceplate, $3\frac{1}{4}$" diam. ...	3 -
Prong Chuck for wood	2 -
Set of 3 Turning Tools	1 6
Set of 6 Turning Tools	3 -
Round Belt, per ft. ...	3d.
Compound Slide-rest ...	9 6
Wood Stand with Flywheel, Treadle and Belt ...	£1 4s.

F. W. PORTASS, 83a, Sellers Street, SHEFFIELD.

No. 106B.

5-in centre by 5 ft. "Star" Screw-Cutting Engine Lathe.

Accurate Lathes for Small Work.

"**STAR**" Screw-cutting Engine Lathes ($4\frac{1}{2}$ and $5\frac{1}{2}$-inch centres) are used in large and increasing numbers for light machine-shop and tool-room service, for model making, electrical and repair work, technical schools, &c. Size and price give them advantages over larger lathes on all work within their capacities.

We also make **FOOT POWER LATHES, BENCH LATHES, SPEED LATHES, WOOD-TURNING LATHES** and **HEAVY ENGINE LATHES.** Send for Catalog "B."

THE SENECA FALLS MFG. CO.,
560, Water Street,
SENECA FALLS, N.Y., U.S.A.

"THE NEW RANDA" 3IN. Sliding Surfacing SCREWCUTTING LATHE

AT THE REMARKABLE PRICE OF £4 : 4 : 0

OR 11 MONTHLY PAYMENTS OF 8/6

THE "RANDA" 3" S.C. LATHE IS MADE BY THE FINEST SMALL LATHE MAKER IN THE WORLD

Headstock bearings are parallel with split gun-metal bushes, having means for adjustment. The mandrel is hollow to admit ⅜ in. bar and for No. 1 Morse taper centres. The tailstock has a hollow square-thread barrel wheel operated and fitted with locking screw. A T-slotted swivelling cross-slide is mounted on saddle operated by square-thread screw and wheel.

The lathe is thoroughly practical in every detail and there is no hesitation in claiming it to be the finest too ever produced at the price and must not be confused with lathes where accuracy and workmanship is sacrificed, or with those offered as clearing lines or so-called bargains.

DIMENSIONS.

Height of Centres	**3 in.**
Distance between Centres ..	**$12\frac{1}{2}$ in.**
Height of Centres from Gap ..	**$4\frac{1}{2}$ in.**
Headstock Mandrel admits ..	**⅜ in.**
Headstock Pulley for Flat Belt.	**3 Speeds.**
BACK GEAR if required, 15/- extra.	
No. 1 Morse Taper Centres.	
Faceplate	**$4\frac{3}{8}$ in.**
Guide Screw	**8 t.p.i.**
Overall Length	**30 in.**
Hollow Tailstock Barrel. Change Wheels, 20, 20, 25, 30, 35, 40, 45, 50, 55, 60.	

ROSS & ALEXANDER, LTD. 165, **BISHOPSGATE, LONDON, E.C.2**

ROSS & ALEXANDER, LTD., Offer You

SOMETHING TO TALK ABOUT!!

FULL RANGE OF ACCESSORIES

"THE NEW RANDA"

3IN. Sliding Surfacing SCREWCUTTING LATHE

The "RANDA" 3" S.C. Lathe is made by the Finest Small Lathe Maker in the World

DIMENSIONS.

Height of Centres ...	3 in.	Faceplate ...	4 3/8 in.
Distance between Centres	12 1/2 in.	Guide Screw	8 T.P.I.
Height of Centres from Gap	4 1/2 in.	Overall Length	30 in.
Headstock Mandrel admits	3/8 in.	Hollow Tailstock Barrel.	
Headstock Pulley for Flat Belt.	3 Speeds.	Change Wheels 20, 25, 30, 35, 40, 45, 50, 55, 60.	
No. 1 Morse Taper Centres.			

PRICE

£4:4:0

Or 11 Monthly Payments of 8/6

ROSS & ALEXANDER, LTD.
165, BISHOPSGATE, LONDON, E.C.2

FOOT & POWER AND TURRET LATHES

PLANERS, SHAPERS, AND DRILL PRESSES. CHUCKS, TOOLS, DOGS.

SHEPARD LATHE CO.,
A 131 W. 2nd St., CINCINNATI, O., U.S.A.
Postage to United States, 2½d. ¼ oz.

STANHOPE 3-in. LATHES.
PLAIN, BACK GEARED OR SCREWCUTTING.
Bench or Treadle. 24" Straight, or 24" or 30" Gap Beds. Machined Parts, Castings, Compound Rests, Foot Motors, etc.
Stamp for List. State Requirements.

The Stanhope Engineering Co.,
Henry Street, Horton Lane, BRADFORD.

FOR THIS MONTH ONLY we will allow 10 per cent. off these prices, with the exception of Aluminium

SHEET BRASS.—1/64", 1s. 7d.; 1/32", 2s. 9d.; 1/16", 5s. 8d.; 3/32", 8s. 6d.; 1/8", 11s. 3d.; 3/16", 17s.; 1/4", 22/- (all per sq. ft.).

SHEET COPPER.—1/64", 1s. 9d.; 1/32", 3s.; 1/16", 6s.; 3/32", 9s.; 1/8", 11s. 9d.; 3/16", 18s.; 1/4", 20s. (all per sq. ft.).

SHEET ALUMINIUM.—1/64", 1s. 7d.; 1/32", 2s. 9d.; 1/16", 5s. 6d.; 3/32", 8s. 3d.; 1/8", 11s. 3d.; 3/16", 17s.; 1/4", 22s. 6d. (all per sq. ft.).

SHEET STEEL.—1/64", 7d.; 1/32", 1s. 1d.; 1/16", 2s.; 3/32", 3s. 3d.; 1/8", 4s. 3d.; 3/16", 6s. 9d.; 1/4", 8s. 9d. (all per sq. ft.).

BRASS RODS, Round.—1/16", 3/32", 1d.; 1/8", 1½d.; 5/32", 2½d.; 3/16" 3d.; 7/32", 4d.; 1/4", 5d.; 9/32", 6d.; 5/16", 7d.; 3/8", 8d.; 7/16", 10d.; 1/2", 1s.; 5/8", 1s. 9d.; 3/4", 2s. 9d.; 7/8", 3s. 6d.; 1", 4s. 2d.; 1 1/4", 6s. 9d.; 1 1/2", 8s. 6d.; 1 3/4", 11s. 9d.; 2", 14s. (all per ft.).

BRASS RODS, Square and Hexagon.—1/8", 2d.; 5/32", 3d.; 3/16", 4½d.; 1/4", 6d.; 5/16", 9d.; 3/8", 11d.; 7/16", 1s. 2d.; 1/2", 1s. 6d.; 5/8", 2s. 3d.; 3/4", 3s. 6d.; 1", 5s.; 1 1/4", 9s.; 1 1/2", 10s. 9d.; 1 3/4", 15s. 3d.; 2", 18s.
Larger sizes in all the above quoted for (all per ft.).

COPPER RODS.—Round, Square, and Flat, Brass, plus 50 per cent.

ANGLE STEEL.—1/2"×1/8", 4d.; 3/4"×1/8", 6d.; 1"×1/8", 8d.; 1"×3/16", 1s.; 1 1/2"×3/16", 1s. 6d.; 2"×1/4", 2s.; 2"×3/8", 3s.; 2 1/2"×3/8", 4s.; 3"×3/8", 4s. 6d.; 3"×1/2", 5s. 9d.; 3 1/2"×1/2", 7s.; 4"×1/2", 7s. 9d. (all per ft.).

COPPER TUBES.—3/32", 3½d.; 1/8", 4d.; 5/32", 5d.; 3/16", 6d.; 1/4", 7d.; 5/16", 8d.; 3/8", 10d.; 7/16", 1s.; 1/2", 1s. 3d.; 5/8", 1s. 6d.; 3/4", 1s. 10d.; 7/8", 2s.; 1", 2s. 3d.; 1 1/4", 2s. 6d.; 1 1/2", 2s. 9d.; 1 3/4", 3s.; 2", 3s. 6d.; 2 1/4", 4s.; 2 1/2", 5s. 3d.; 2 3/4", 6s.; 3", 6s. 9d.; 3 1/2", 8s. 6d.; 4", 10s.; 4 1/2", 11s. 3d.; 5", 13s. 6d.; 6", 16s. 3d. (all per ft.).
Thicknesses up to 1/4" are 20/21 Gauge, larger 16 Gauge.

BRASS TUBES.—Same price as Copper Tubes, less 20 per cent.

ALUMINIUM TUBES.—Same price as Copper Tubes.

BRASS SQUARE TUBES.—Up to 1" sq., same price as Copper.

BRASS FLAT RODS.—

1/4"×1/32",	2d.	1/4"×1/16",	2d.	1/4"×3/32",	3d.	1/4"×1/8",	4d.
3/8"×1/32",	2d.	3/8"×1/16",	3d.	3/8"×3/32",	4d.	3/8"×1/8",	6d.
1/2"×1/32",	3d.	1/2"×1/16",	4d.	1/2"×3/32",	5d.	1/2"×1/8",	8d.
3/4"×1/32",	4d.	5/8"×1/16",	5d.	5/8"×3/32",	6d.	5/8"×1/8",	9d.
1"×1/32",	6d.	3/4"×1/16",	6d.	3/4"×3/32",	7d.	3/4"×1/8",	1/-
		1"×1/16",	8d.	1"×3/32",	9d.	1"×1/8",	1/-
Wider, ½d. per 1/8".		Wider, 1d. per 1/8".		Wider, 1½d. per 1/8".		Wider, 2d. per 1/8".	

1/4"×3/16"	5d.	3/8"×1/4",	11d.	3/8"×5/16",	1/3	1/2"×3/8",	2/-
3/8"×3/16",	8d.	1/2"×1/4",	1/3	1/2"×5/16",	1/8	5/8"×3/8",	2/7
1/2"×3/16",	10d.	5/8"×1/4",	1/7	5/8"×5/16",	2/-	3/4"×3/8",	3/-
5/8"×3/16",	1/1	3/4"×1/4",	1/10	3/4"×5/16",	2/4	1"×3/8",	4/-
3/4"×3/16",	1/4	1"×1/4",	2/2	1"×5/16",	2/9	1 1/4"×1/2",	5/9
1"×3/16",	1/7	1 1/4"×1/4",	2/9	3/4"×1/2",	3/7	1 1/2"×1/2",	6/9
Wider, 2½d. per 1/8".		Wider, 3d. per 1/8".		1"×1/2",	4/9	2"×1/2",	9/6

BRIGHT MILD STEEL RODS.—Round, Square, Hexagon, and Flats. Price of Brass, less 50 per cent.

2d. will bring you our Price List. Postage Extra.

PUGH & DAVIS, 47, HERBERT STREET, NEW NORTH ROAD, SHOREDITCH, LONDON, N.1.

Drill Blocks and Clamp.

THE DRILL BLOCKS ARE FURNISHED IN PAIRS. THE SIZE OF EACH IS 2 inches by $1\frac{1}{2}$ inches.

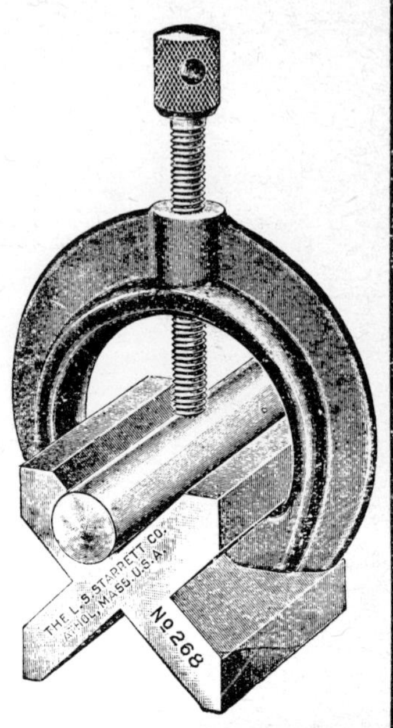

The Clamp will hold a round piece up to $1\frac{1}{8}$ inch diameter firmly in the groove of the blocks. For prick punching, drilling, or laying out a series of holes before and while being drilled.

ASK YOUR TOOL DEALER FOR PRICES.

Have you had a copy of our new Catalogue No. 19 E.M.? Your Tool Dealer will hand you this book free of charge, or we will send on receipt of 3d. in Stamps to pay Postage.

The L. S. STARRETT Co.,

36-37, Upper Thames St., London, E.C.

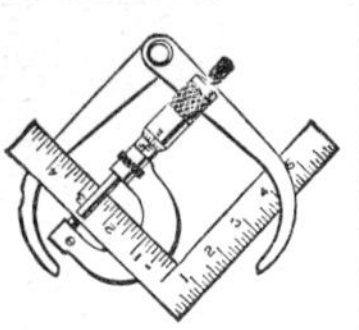

ACCURACY

IS THE FIRST CONSIDERATION IN THE MANUFACTURE OF

Starrett Micrometers

STARRETT'S No. 2A MICROMETER will take all measurement from 0 to 2 inches. As shown below it will do the work of a regular inch micrometer. The Attachment may be removed by turning the thumbscrew slightly. You then have a regular 2-inch micrometer.

Two Micrometers in One, and Both Accurate.

Our New No. 19 E.M. Catalogue has 275 pages showing the Starrett line up to date.

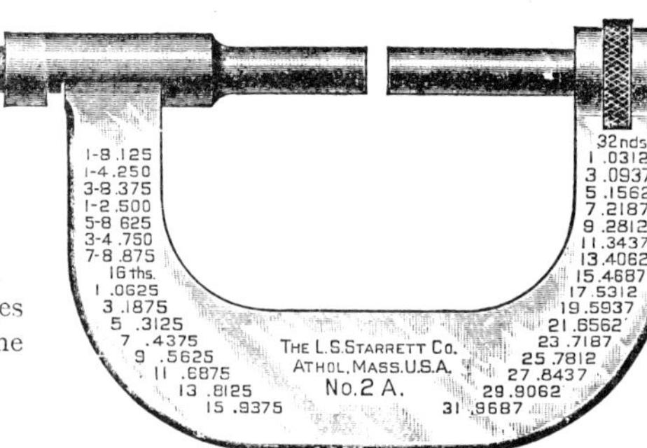

In our New Catalogue you will find listed a complete line of Micrometers.

Ask for Starrett Catalogue No. 19 E.M.

It is free of any Tool Dealer.

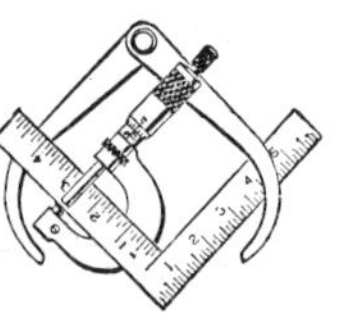

Or send us 3d. in stamps and we will send the book on to you carriage paid.

THE L. S. STARRETT CO.,

36-37, UPPER THAMES STREET, LONDON, E.C.

The IMPROVED "ZYTO" T.R. LATHE

with TUMBLER REVERSE.

The Introduction of the Improved "Zyto" T.R. Back-geared Screw-cutting Lathe with **TUMBLER REVERSE** forms a notable example of the production of a really high-grade machine tool at an exceptionally low price.

3" Centre Sliding, Milling, Surfacing, and Screwcutting Lathe—WITH TUMBLER REVERSE

DIMENSIONS:

Height of centres......	3"	Hole through mandrel and tailstock	⅜"
Length between centres	12"	Centres, Morse taper No. 1	
Height from gap	3¾"	Lead Screws ...	10 t.p.i.
Height from saddle...	2½"	Change Wheels	11 ,,

PRICES:

	£	s.	d.
Improved "ZYTO" T.R. Back-geared Screw-cutting Lathe	8	2	6
Complete Lathe Stand, with chip tray and heavy 3-speed Flywheel, pedal ...	4	4	0
Compound Rest Model ... extra	1	10	0

(No extra charge for deferred payments.)

Catalogue of SPECIAL OFFERS Free and post free.
Complete Catalogue (257 pages), fully illustrated, 2/-.
(This amount may be deducted from first order of 10/-.)

S. TYZACK & SON, Ltd., 341-345, Old Street, Shoreditch, LONDON, E.C.1.

THE ESSENTIAL POINTS of an ACCURATE and RELIABLE LATHE

Dimensions: Height of centres, 3"; Length between centres, 12½" (also supplied 18" between centres, 30/- extra); Height from gap, 4½"; Height from saddle, 2½"; Hole through mandrel and tail-stock, ⅜"; centres, Morse Taper No. 1; Lead Screw, ⅝", 8 T.P.I.; Change Wheels, 10.

S. TYZACK & SON, Ltd., 341-345, Old St., London, E.C.1

Ball Bearing Thrust. The provision of a ball race is necessary to take the thrust of the mandrel when using the lathe as a drilling machine.

Tumbler Reverse. The importance of this feature is obvious in connection with screw cutting and the use of self-acting traverse.

Rigidity. The bed and headstock casting must be of ample proportions to resist any tendency to distortion or "chattering" when taking heavy cuts.

Gunmetal Headstock Bearings are essential for the Mandrel and these must be adjustable.

All of the above features are found in the Improved "ZYTO" Lathe —a really high grade machine tool obtainable at an exceptionally low price and on the easiest of deferred terms.

First payment 12/6 and 15 monthly payments of 10/-

PRICES.

	£	s.	d.
Improved "ZYTO" T.R. Back-geared, Screw-cutting Lathe	8	2	6
Complete Lathe Stand, with chip tray, heavy 3-speed Flywheel and Treadle	4	4	0
Compound Rest, if required, extra	1	10	0

All less 10% cash

LATHES.

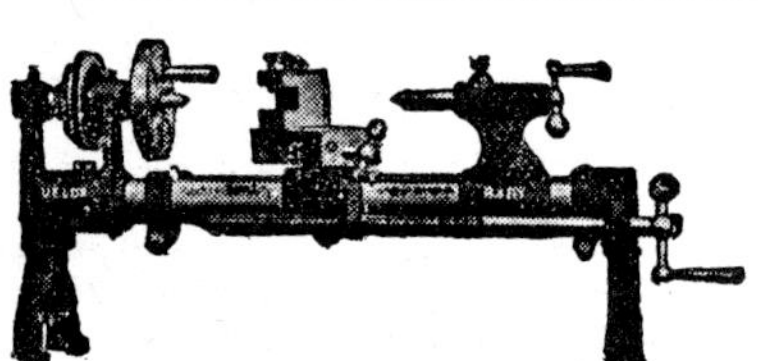

The 1922 Velox 2½" centre Baby Super Lathe is designed for Turning, Boring, Milling, and Gear-cutting.
Price Complete £4 : 18 : 6.
As Plain Lathe only, £2 : 7 : 6.

Further particulars, Stamp.

THE VELOX Tool Manufacturing, Electrical, and Wireless Engineering Co.,
255, Westgate Rd., Newcastle-on-Tyne.

2½" Screw-cutting Lathe, £7.

With Back-gear and Gap-bed, Machine-cut Gears.

Above Lathe, but without Back-gear	£5 10s.
Above Lathe, on heavy Treadle, extra ...	£3.
12 Slide-rest Tools and Toolholder ...	6/-
3 Slide-rest Tools and Boring Bar, ...	3/6.

Particulars, Stamp, please.

THE UNION TOOL CO.,
89, ST. DONATTS ROAD, LONDON, S.E.14.

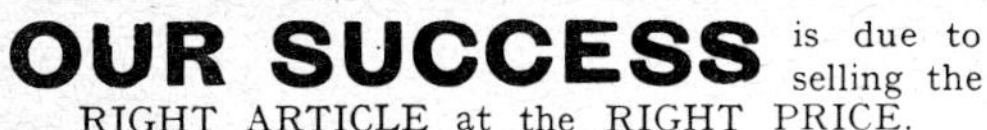

OUR SUCCESS is due to selling the RIGHT ARTICLE at the RIGHT PRICE.

2½-in. Back-geared Screw-cutting Lathe, Gap Bed, Machine-Cut Gears, **£7 0 0.**

2½-in. Back-geared Screw-cutting Lathe, on Treadle, **£10 0 0.**

2-in. Lathe with Cast Iron Gap Bed and Slide Rest, **£1 15 0.**

See what our Customers think of our Goods——in our New Illustrated Catalogue of Lathes, Tools, etc. Post Free, 3d.

**THE UNION TOOL CO.,
89, St. Donatt's Road, London, S.E.14.**

LATHES: "The Ideal" 3″ CENTRE, S.C., B.G.

See Stand No. 40 at the "M.E." Exhibition, for

New 3½″ Model S.C., B.G., ¾″ Hollow Spindle.

Particulars.——Stamp Please.

WILLIMOTT & SON, Nevilles Factory, CHILWELL, Notts.

"WREN" Hollow Tailstock DIE-HOLDERS.

An Indispensable Accessory for every lathe.

SAVES TIME. ENSURES ACCURACY.

SIZES.			PRICE
No. 1.	For 13/16″ diameter Dies. No. 1 M.T. Shank (Hole 5/16″)	...	**7/6**
No. 2.	For 1″ diameter Dies. No. 1 M.T. Shank (Hole 5/16″)	...	**12/6**
No. 3.	For 1″ diameter Dies. No. 2 M.T. Shank (Hole 7/16″)	...	**15/6**

Size 1, Hexagon Steel. Sizes 2 and 3, Round Steel. With Flats.

No. 1 size, for **WADE LATHES, 8/6.**

DIES to suit above Holders.—13/16″ diam. Whit., **1/-** ea.; 1″ diam. Whit., **1/9** ea.; Fine thread Dies, **1/3** and **2/3** ea.

MADE BY: *Also obtainable from High-class Tool Dealers.*

**The WREN TOOL CO.,
174, Sixth Avenue, Manor Park, LONDON, E.12.**

FOUR-JAW CHUCKS, 14/-!

The Most Amazing Chuck Offer ever made.
We stock these ready screwed to fit your Mandrel Nose.
4″ diam. Chucks, 14/- (screwed up to ¾″); 4½″ diam. (screwed up to 1″), 16/-; 5¾″ diam. (screwed up to 1¼″), 25/-. All post paid. Order To-day, while we have stock!
3″ Self-Centring Chucks, 2 sets jaws, 24/-; Finished Backplates, 5/-.

THE WREN TOOL CO., 236, Sumner Rd., London, S.E.15.

MILLING ATTACHMENT and VICE for the LATHE.

To suit all Lathes from 3½″ to 6″ centres. A most useful addition to any Lathe.

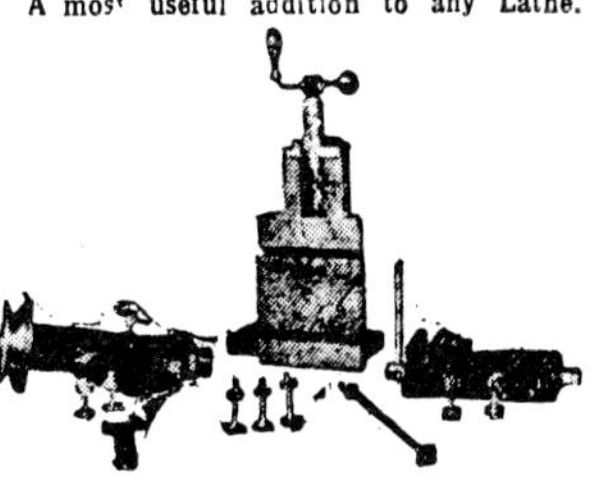

Vertical Slide and Table	...	£2 17 11
Swivel Index	...	£2 1 9
Vice	...	£1 9 3
Set of Cutters	...	£1 14 11
Total	...	£8 3 10

Carriage paid United Kingdom.
Foreign Customers add 10% for postage.

WHEELER MANFG. Co., Ltd.,
Trench Crossing, near WELLINGTON, Salop.

"IDEAL" LATHES.

**3″ S.C.B.G. from £7 : 0 : 0.
3½″ S.C.B.G. „ £7 : 18 : 6.**

J. WILLIMOTT & SON,
Nevilles Factory, CHILWELL, Notts.

The "WALRAM" Back Gear ATTACHMENT for 4″ "Drummond" Lathe, is

**SIMPLE! CHEAP!
FITTED in a few MINUTES.**

The best Back Gear Attachment yet devised for the "Drummond" Lathe.
Send for full particulars:

**WALRAM ENGINEERING CO.,
25, Heyes Lane, TIMPERLEY, MANCHESTER.**

PORTASS 2⅝″ Cen. LATHES.

55/- With lever tailstock **45/-**. *Deferred Payments 5% extra.* **6/- WITH ORDER.** 9 Monthly Payments.

WRITE DEPT. *M.E.*
**GREGORY & TAYLOR, Ltd.,
14, Mount Pleasant Rd.,
SHEFFIELD.**